SCI FI
DECLASSIFIED

The Roswell Dig Diaries

SCI FI
DECLASSIFIED

The Roswell Dig Diaries

A SCI FI Channel Book

Including Contributions from
William H. Doleman, Ph.D.,
Thomas J. Carey, and Donald R. Schmitt

Foreword by New Mexico Governor Bill Richardson

Edited by Mike McAvennie

POCKET BOOKS

New York   London   Toronto   Sydney

POCKET BOOKS, a division of Simon & Schuster, Inc.
1230 Avenue of the Americas, New York, NY 10020

ISBN: 0-7434-8612-9

First Pocket Books trade paperback edition July 2004

10 9 8 7 6 5 4 3 2 1

POCKET and colophon are registered trademarks of
Simon & Schuster, Inc.

All photos Babak Dowlatshahi/InsightPhotos.com/SCI FI Channel
Courtesy of SCI FI Channel

Manufactured in the United States of America

Designed by Joseph Rutt

For information regarding special discounts for bulk purchases, please contact
Simon & Schuster Special Sales at 1-800-456-6798 or business@simonandschuster.com

CONTENTS

FOREWORD

To many Americans and many people around the world, Roswell, New Mexico, carries a special meaning. For a wide variety of reasons, Roswell represents a place where something—an extraterrestrial spaceship, according to some, a sophisticated weather balloon to others—crashed in July 1947.

The mystery surrounding this crash has never been adequately explained—not by independent investigators, and not by the U.S. government. In 2002, using the latest high-tech tools, scientists from the University of New Mexico initiated an archeological dig sponsored by the SCI FI Channel to help determine what crashed into the desert near Roswell.

Using all of the tools and capabilities of modern science, we need to get to the bottom of something that has perplexed millions of Americans, and others, for more than fifty-seven years. The mystery surrounding the Roswell crash has become a fixture of American culture. It has been the focus of dozens of books, documentaries, TV specials, and movies. There are as many theories as there are official explanations.

Clearly, it would help everyone if the U.S. government disclosed everything it knows. Openness is our best policy. The American people can handle the truth—no matter how bizarre or mundane. And after more than fifty-five years, every government document on the Roswell incident should be released to the public. With full disclosure and our best scientific investigation, we should be able to find out what happened on that fateful day in July 1947.

The Honorable Bill Richardson
Office of the Governor
Santa Fe, New Mexico
June 2004

INTRODUCTION: DIGGING UP THE TRUTH

by Thomas J. Carey and Donald R. Schmitt

The alleged Roswell UFO crash of 1947 remains one of the great unsolved mysteries of the twentieth century, if not in the annals of history. Something truly extraordinary crashed on an isolated ranch in central New Mexico, and after fifty-plus years the high desert country still protects its secrets. We know that fact better than anyone, thanks to what transpired within days of the initial event.

The U.S. military went to extreme lengths to conceal its own concern as well as the true nature of the strange wreckage. An official press release, dated July 8, 1947, verified the Army Air Force's recovery of a flying disc, as UFOs were then called, near the sleepy desert town of Roswell, New Mexico. Later the same day, that astonishing claim was retracted by the military, and the unknown object was said to have been nothing more than a common weather balloon device. And so the secret was kept intact as the prevailing desert winds resumed inexorably over the silent, vacant ranchland.

But behind the scenes, the government was using all available means, legal and illegal, to quash media interest in "the story of the millennium." The crash site was cordoned off from all outsiders, including the press, and every shred of evidence was retrieved. But time has a way of mitigating such extreme efforts, and the ultimate proof still remains. That is, of course, if one is willing to look for it.

The cover-up was executed with such dispatch and precision that the Roswell Incident,* as it is now generally referred to, was thought to have

---

* Derived from the title of the initial book on the subject, *The Roswell Incident,* by Charles Berlitz and William Moore (New York: Grosset & Dunlap, 1980).

been effectively contained and relegated to the "Silly Season" file. But no secret, especially a government secret, is safe these days, and Roswell is no exception. This fact became quite evident when former Roswell base intelligence officer Major Jesse A. Marcel decided to reveal his role in those long-ago events. "It was nothing made on this earth," he declared. The truth was beginning to emerge.

This was the very first indication that the original press release stating that a crashed UFO had been "captured" was the correct one. Additional witnesses would soon be located or come forward, and a new version of what transpired at Roswell in 1947 began to peel away the layers of a cover-up. One of the incident's most puzzling aspects was why the military in Roswell would continue to detain ranch foreman W. W. "Mack" Brazel for days after the balloon story had been accepted by the press as the alternative explanation. After all, it was Brazel who had originally alerted Marcel and others to the massive amount of unusual wreckage on a remote ranch seventy-five miles northwest of Roswell. And it was his son, Bill Brazel Jr., who, upon learning of Mack's fate, rushed from his home and new bride in Albuquerque to the ranch and filled in for his absent father. Not only would Bill Jr. discover the heavy vehicle tracks left behind by the military during the cleanup, but he would also become a firsthand eyewitness to one of the most striking pieces of evidence that still intrigues researchers of the Roswell story today: a 500-foot gouge, starting at the upper northwest pinnacle of the debris field site, where something had impacted the surface of the hard desert terrain.

After his father's return, Bill Brazel Jr. would continue to help out on the ranch. During this period, he fortuitously found several small pieces of the wreckage scattered nearby, but not buried. His descriptions of what he found were consistent with those of Major Marcel. Unfortunately, when the military found out two years later about the "scraps" that Bill Jr. was keeping in a cigar box, air force officers from Roswell went to his home and confiscated everything. Nevertheless, the fact that pieces of wreckage were found after the military recovery operation at the crash site implies that more debris may still exist around the location. Of course, it is one thing to deduce that there are artifacts from 1947 still at the crash site; it is quite another matter to find them.

As with the case against the UFO phenomenon in general, critics cite the lack of hard physical evidence in the Roswell, New Mexico, case as proof that nothing of an extraterrestrial nature occurred there in July 1947, as witnesses and researchers contend. Although witness testimony is considered evidence in courts of law throughout the land, the testimonies from individuals bearing witness to the extraordinary events, now numbering into the hundreds, are not considered sufficient as proof by skeptics, debunkers, academics, and most elite media. Incontrovertible physical evidence, they say, must be obtained, and good science must be brought to bear upon it before definitive conclusions can be made. These same worthies, however, uncritically accept the air force's balloon explanation as fact, without a smidgen of evidence—let alone proof—to support it.

Most UFO researchers, scientists, and academics alike, agree that the subject of UFOs, unlike other fields of study, does not readily lend itself to the application of the so-called scientific method for studying the phenomenon in order to produce verifiable results that can be tested and replicated (requirements for any conclusions to be drawn from such studies). One aspect of the Roswell case, however, does lend itself to the implementation of scientific methods of scrutiny—namely, allegations that physical artifacts were recovered from the crash. To date, all such claimed artifacts, when subjected to testing, have been shown to be either definitely of earthly origin (as in the case of a piece of Japanese jewelry a few years ago) or not demonstrably otherworldly in nature. Is waiting for someone to show up on our doorstep with the Holy Grail of Roswell the best we can do in our quest to find the answer?

Over the twenty-five years that the Roswell case has been investigated, researchers have suggested that there may have been as many as three separate but related crash sites involved in the event. Not all agree, however, on their locations, with the notable exception of one site. Even the air force agrees on this one site, though it disagrees that it was a UFO that crashed there. This site is located on the former J. B. Foster Ranch, where Mack Brazel found a large sheep pasture covered with pieces of strange wreckage. The air force does not dispute the location of this site, and it agrees that something crashed there in 1947, but it disputes what it was, choosing to believe instead that Brazel found the remains of a high-altitude balloon array and acoustic device used in a failed attempt to detect the expected detonation of the Soviet Union's first atomic bomb.

What was it then—a UFO, as civilian investigators suggest, or a balloon and tinfoil device, as per the official position of the air force today? These are the only choices now available to us, after years of study and controversy. Although the site was "vacuumed" back in 1947, testimony by the late Bill Brazel Jr. and others suggest that the military did not retrieve all of the debris. Wind action may have carried pieces of the debris far from their original area of concentration to more outlying areas, and burrowing animals may have carried pieces into their subterranean nests. Whatever it was, it is not too far-fetched to think that several, or even many, pieces remain buried within just a few feet of the surface of the desert floor and, therefore, might be obtained using standard archeological field methodology. If this were successful, we would finally have the answer once and for all.

In 1989 a small band of researchers from the J. Allen Hynek Center for UFO Studies in Chicago attempted to conduct a preliminary archeological survey of the acknowledged (Brazel) debris field site associated with the 1947 Roswell events. Meagerly funded and short on trained personnel, this underequipped "dig" turned up a rusted tin can and a single shell casing for its efforts. Understandably, no report of this endeavor was ever written. A return to the site to conduct a follow-up excavation was planned, but the necessary funding never materialized; the idea never came to fruition and was forgotten.

In May 1998, the two of us joined forces to continue a proactive investigation of the case. In 1999 we approached the University of New Mexico's Office of Contract Archeology in Albuquerque about the feasibility of conducting a full-scale historical archeological "dig" at the Foster Ranch crash site in Lincoln County, New Mexico, for the purpose of unearthing whatever might lie buried just beneath the surface of the earth there. To our gratification, the archeologists at UNM concurred that it could be done in a normal manner with standard archeological field techniques that would be used for any dig in that locale. Everyone agreed that the dig should be undertaken, given its historical import to the state of New Mexico as well as its potential payoff—the definitive answer to one of the great mysteries of all time, and the possibility of proof that we are not alone in the universe.

Once in hand, we shopped the dig proposal around for several years to potentially interested sources of funding, without success. In the spring of 2002 we were approached by the SCI FI Channel's Larry Landsman about working together on a project connected to Roswell. A quick trip to Roswell with Larry in June 2002 solidified SCI FI's commitment to an excavation, and the dig commenced three months later.

Tom Carey
Huntingdon Valley, Pennsylvania

Don Schmitt
Hubertus, Wisconsin

THE STORY SO FAR . . .

A Time Line

Over the course of their work, UFO investigators Thomas Carey and Donald Schmitt compiled a historical timeline of events that comprise the "Roswell Incident." This is the result.

June 24, 1947

While searching for a downed military aircraft, Boise, Idaho, businessman Kenneth Arnold spots nine silver, crescent-shaped objects traveling at speeds of 1,200 mph near Mount Rainier, Washington. This ushers in the modern age of unidentified flying objects (UFOs), and for the next two weeks, flying saucers make the front pages of newspapers across the country.

July 1, 1947

Sherman Campbell discovers weather balloon and radar target debris on his farm near Circleville, Ohio.

July 4, 1947 (date uncertain)

10:30 P.M.: Several ranchers north of Roswell report seeing a fiery object arc toward the ground.

Late evening: W. W. "Mack" Brazel, ranch foreman at the J. B. Foster sheep ranch (located thirty-five miles south-southeast of Corona, seventy-five miles northwest of Roswell), hears an explosion during a severe thunder and lightning storm.

Roswell residents William Woody and Bob Wolf report seeing a fiery object descend north of town. Woody believes the object landed in an area southwest of where the Corona road crosses Highway 285.

While scanning the night sky atop St. Mary's Hospital in Roswell, several nuns see a flaming object fall to the north. They record the event in their nightly logs.

July 5, 1947 (date uncertain)

Morning: While checking his sheep with seven-year-old Dee Proctor, Mack Brazel discovers a large amount of strange wreckage covering most of Hines Pasture. Brazel returns the Proctor boy to his parents, Floyd and Loretta, then shows them a small piece of the wreckage. Declining the rancher's urgings to accompany him back for another look, the Proctors suggest the debris may be from a flying saucer, as they understand a newspaper somewhere has posted a reward for parts of one.

Brazel hauls and stores several large pieces of wreckage inside a cattle shed about 100 feet from Hines House, a small bunkhouse on the J. B. Foster Ranch that's approximately three miles from the debris-field site.

Afternoon–evening: Seeking opinions, Brazel visits his neighbors, the Lyman Stricklands, then Wade's Bar in Corona, and passes around a piece of the wreckage. A relative advises that he take it to the air base in Roswell.

July 6, 1947

Early morning: Brazel notices birds of prey circling off in the distance and a faint, foul odor in the air. He saddles up and rides over to a low bluff 2.5 miles east of the Hines Draw debris field. Ascending to the top of the bluff, Brazel finds more wreckage, plus three small dead bodies.

Noon: Loading his pickup truck with two cardboard boxes full of debris, Brazel drives to the Chaves County sheriff's office in downtown Roswell, where he informs Sheriff George Wilcox of his find. Frank Joyce, an announcer for Roswell radio station KGFL, calls the sheriff and asks for news items to air on his show. At Wilcox's suggestion, Brazel tells Joyce about finding the wreckage and bodies of "little people." When Joyce skeptically suggests the bodies may be monkeys from some air force experimental rocket, the enraged rancher yells, "They're not monkeys, goddammit! They're not human!" and slams down the phone. Sheriff Wilcox calls Roswell Army Air Field (RAAF) and is connected with the 509th Bomb Group's intelligence officer, Major Jesse A. Marcel, who promises to come after lunch.

1:00–2:00 P.M.: After interviewing Brazel at the sheriff's office, Marcel takes one of the boxes of wreckage back to the RAAF base to show his commanding officer, Colonel William H. "Butch" Blanchard. The second box remains at the sheriff's office.

2:00–3:00 P.M.: Colonel Blanchard orders Marcel to follow Brazel out to the debris site. Marcel contacts CIC captain Sheridan Cavitt, who drives a Jeep Carry-All while Marcel follows the rancher's pickup truck in his own '42 Buick convertible. Blanchard alerts his next higher command to the find.

Late afternoon: SAC Command orders Blanchard to immediately send the secured box of wreckage from Fort Worth Army Air Field to Washington, D.C., via bomber/transport.

Early evening: Too late in the day to accomplish anything outside, Brazel, Marcel, and Cavitt spend the night at the Hines house and inspect the wreckage the rancher had stored in the cattle shed.

Two deputies dispatched by Sheriff Wilcox report finding two areas of blackened ground to the north and west of Roswell. Growing darkness prevents further investigation.

July 7, 1947

Early morning: Marcel, Cavitt, and Brazel head for the Hines Pasture debris site, which Marcel would later describe as approximately three-quarters of a mile long, 200 to 300 feet wide, and covered with small pieces of extremely thin aluminum-like foil. He would also later opine the foil's provenance as being "not from this earth." They load the Jeep Carry-All with as much debris as it will hold.

Morning: A group of archeologists traveling cross-country and looking for signs of Paleo-Indians discover the crash site forty miles north of Roswell. A few local ranchers also come across the site, and Roswell's sheriff's office and fire department are quickly notified.

Sheriff Wilcox contacts the RAAF, then either follows or leads the fire engine to the crash site.

Ranchers and archeologists are warned not to discuss what they've seen, and the latter are taken back to the RAAF base, where they're detained and debriefed. The sheriff's office and fire department personnel are appealed to under the veil of "patriotism," but ultimately are threatened with their lives and those of their family members.

Late morning: Walt Whitmore Sr. and Jud Roberts, owners of Roswell radio station KGFL, arrive at the J. B. Foster Ranch, then drive Brazel back to Roswell for the purpose of interviewing him. A wire-recorded interview of his story is made for the purpose of airing it on the station as a "scoop." Brazel is hidden overnight at Whitmore's house.

Afternoon: Returning to the RAAF base with a Jeep Carry-All full of wreckage, Cavitt reports to Blanchard and begins writing a report for CIC Command. Marcel drives back to the debris field and spends the rest of the day gathering additional wreckage. He does not visit the "Dee Proctor site" three miles east of the Foster Ranch debris field, where Brazel earlier informed him that "something else" could be found.

At the crash site north of Roswell, the recovery process begins. Military police (MP) units are stationed along the roads north of town, blocking access to the site west of Highway 285. More MPs are added to those ringing the crashed spacecraft.

Evening: Removed from the crash site, the recovered bodies are taken in a field ambulance to Hangar P-3 (known today as Building 84). Assigned to guard the ambulance, Staff Sergeant Melvin Brown, a cook with the 509th's "K" Squadron, peeks under the tarp and sees the bodies (an experience he mentions only on his deathbed, many years later). Packed overnight in dry ice, the bodies are placed in a side room within the hangar.

Late evening: Marcel drives to Roswell after filling his '42 Buick with what he believes is wreckage from a flying saucer.

July 8, 1947

2:00 A.M.: Marcel heads to his house in Roswell, where he wakes his wife and son and shows them the extraordinary wreckage ("pieces of a flying saucer," according to his son). As he places several parts down on the kitchen floor and tries piecing them together, his son, Jesse Jr., notices what appears to be unrecognizable symbols or writing along the inner surface of one of several small rods that are shaped like an I-beam in cross section. An hour or two later, Marcel reloads his '42 Buick and drives to the base.

7:00 A.M.: An emergency staff meeting is held in Colonel Blanchard's office to discuss the situation. General Roger M. Ramey, commanding officer of the "Mighty" Eighth Air Force and Blanchard's boss, and his chief of staff, Colonel Thomas J. DuBose, fly in from Fort Worth, Texas, and join PIO First Lieutenant Walter Haut, Major Marcel, Captain Cavitt, and Blanchard's regular staff for the meeting. Upon examining the wreckage, they decide to draft and distribute a press release announcing the recovery of a flying saucer, to assuage local rumors of crashed flying saucers and dead aliens running rampant in Roswell. When Marcel mentions the Dee Proctor site, they immediately dispatch a recovery team and extend the cordon along Highway 285. Another team is dispatched to the Foster Ranch debris field site to begin cleanup. To better control the information flow, it is also determined that Brazel be located and detained.

Morning: Johnny McBoyle, a reporter for radio station KSWS, runs into Brazel at a local Roswell coffee shop and hears about the crash. He gets directions from Brazel and heads for the debris field site.

The bodies are flown to Wright Field in Dayton, Ohio. The C-54 flight, piloted by Captain Oliver W. "Pappy" Henderson of the First Air Transport Unit (the "Green Hornets"), stops first at Andrews Air Force Base in Washington, D.C., where the bodies are viewed by selected high-ranking military and civilian personnel.

Mid-morning: Brazel is located at Walt Whitmore's house and taken into custody by a RAAF security detail. For the next week or so, he is quartered at the base guesthouse, just inside the main gate.

Late morning: The bulk of the wreckage brought in by Marcel and Cavitt is boxed in wooden crates, loaded, and flown out on C-54s to various destinations.

Wreckage from the crash site north of Roswell is placed on a flatbed, under a tarp, and driven through side streets to Hangar P-3. Additional security is brought in from White Sands, Alamogordo, and Fort Bliss to help secure the base and control events and personnel.

11:00 A.M.–noon: Under the authority of Colonel Blanchard, 1st Lt. Walter Haut writes up and distributes a brief press release to the effect that the RAAF has captured a downed flying saucer in the "Roswell region." There is no mention of a rancher finding the wreckage or anything about alien bodies, though the release does acknowledge Major Marcel and the fact that the wreckage was being taken out of Roswell and "loaned" to a higher authority. Haut hand-carries the release to local radio stations KGFL and KSWS, plus the *Roswell Daily Record* (an evening newspaper) and *Roswell Morning Dispatch* (a morning paper).

Noon: Apprehended by military security while trying to reach the debris field, Johnny McBoyle is taken to the RAAF base for debriefing. While there, McBoyle observes the recovery operation in progress and attempts to get the scoop on the AP wire by calling his parent station in Albuquerque. Lydia Sleppy receives the call, but as she tries typing the story for transmission, she's cut off by a message that states, "CEASE TRANSMISSION . . . NATIONAL SECURITY MATTER . . . REPEAT . . . CEASE TRANSMISSION . . . NATIONAL SECURITY MATTER." McBoyle is pulled aside by MPs, and his life is threatened. The reporter tells Sleppy, "Forget it. It was nothing," and never discusses the incident again.

KGFL announcer Frank Joyce puts the Haut press release on the United Press (UP) wire, making it a national and international news event.

Early afternoon: Major Marcel boards *Dave's Dream,* a B-29 plane that Colonel Blanchard personally orders to fly to Fort Worth Army Air Field. "FBI types" load several boxes of wreckage (though not the bulk of it recovered from the Foster Ranch) into the bomb bay. Blanchard and First Lieutenant Robert Shirkey, the flight operations officer on duty, witness firsthand the departure of the flight, which is piloted by the deputy base commander, Colonel Payne Jennings, with Master Sergeant Robert Porter acting as crew chief. Throughout the trip, Marcel keeps on his lap a small sampling of the wreckage to personally deliver to General Ramey.

News of the recovery spreads throughout the country, then the world, as the story hits wire services. Phone lines at the base, the sheriff's office, the media, and even the Marcel house are jammed with callers. General Clements McMullen, deputy commander of SAC in Washington, orders Colonel DuBose in Fort Worth to concoct a story that will put out the burgeoning firestorm and kill press interest in the story. General Ramey, having returned to Fort Worth to prepare for a news conference concerning the Roswell events, is already stating that it was a weather balloon, not a flying saucer, that was found.

CIC Captain Sheridan Cavitt takes Master Sergeant Lewis "Bill" Rickett to the crash site north of Roswell to solicit his immediate subordinate's reaction and opinion before completing his report.

Afternoon: Glenn Dennis, an embalmer at the Ballard Funeral Home in Roswell, receives a series of phone calls from the RAAF base mortuary officer, who inquires about the best method of preserving tissue that's been exposed to desert elements. The officer also asks if he has any "children's caskets" in stock, to which Dennis replies that he has one, but can obtain as many as are needed by the close of the following day. Dennis asks if there has been an air accident and offers to drive out to the base, but the officer insists there has been no accident; his inquiries are allegedly information-seeking calls.

Colonel Blanchard—"on leave" from the base, making him unavailable to the press—sets up a command post at the crash site north of Roswell to direct the recovery operation.

New Mexico senator Dennis Chavez calls KGFL station owner Walt Whitmore Sr., warning him not to air the Brazel wire-recorded interview if he wants to remain in business. Someone from the Federal Communications Commission (FCC) also calls Whitmore with the same message, while Frank Joyce receives a threatening call from the Pentagon. Joyce protests that the caller can do nothing to a civilian. Before hanging up, the caller replies, "I'll show you what I can do!"

The first truckloads of wreckage from the Corona debris-field site reach the RAAF base, and are taken to Hangar P-3 for storage.

Three bodies recovered from the Dee Proctor site, along with a moderate amount of wreckage, arrive at the RAAF base hospital via military ambulance.

Friends see Mack Brazel walk down a Roswell street in the company of an armed military security detail. Perhaps also accompanied by Walt Whitmore Sr., Brazel visits and informs the *Roswell Daily Record* that he had found a weather balloon on his ranch, not a flying saucer as had been previously reported. Brazel's photo is taken and released on the AP wire—the first wire photo ever sent out from Roswell.

Dave's Dream lands in Fort Worth, where Marcel brings his box of wreckage to General Ramey's office and places it on his desk. Ramey asks Marcel to follow him into a room with a large wall map and point out exactly where the wreckage was found. Reentering Ramey's office after a brief discussion, Marcel notices that the box is gone, and spread out on the floor are the remains of a rubber weather balloon and a torn-up radar target. J. Bond Johnson, a reporter/photographer for the *Fort Worth Star Telegram*, enters the room and takes photographs of Marcel, Ramey, Colonel DuBose, and Warrant Officer Irving Newton (the base weather officer on duty) posing with the balloon and radar target. Declaring the flight to Wright Field has been canceled, Ramey orders Marcel to return to his duties in Roswell.

Late afternoon: Cleanup continues at both crash sites.

Mack Brazel, still accompanied by the RAAF security detail, visits Frank Joyce at KGFL and gives him the new weather balloon explanation. The surprised announcer asks what is happening, but Brazel insists, "I have to say this. It will go hard on me if I don't." Joyce asks, "What about the 'little green men' you told me about?" to which Brazel responds, "They weren't green," and leaves the station.

A preliminary autopsy of the bodies from the Dee Proctor site is attempted at the RAAF base hospital.

In Roswell, Glenn Dennis is dispatched to transport an airman injured in a motorcycle accident to the RAAF base hospital for treatment. Noticing increased activity and security around the hospital, Dennis passes several field ambulances parked close by the entrance. Looking inside one that has its rear doors open, he sees several large pieces of metallic wreckage ("like burnt steel," he recalls); one piece is shaped like the end of a canoe, with unfamiliar symbols or writing on its surface. Dennis drops off the injured airman inside the hospital, but the beehive of activity and unfamiliar personnel prompts him to ask if there's been a plane crash. He is told first to remain where he's standing, then to leave the building. On his way out, he bumps into a nurse he knows. The nurse, crying and holding a towel over her face, warns him to leave. A redheaded army captain confronts and threatens Dennis, who protests that he's a civilian and cannot be ordered around by the military. The captain warns, "They'll be picking your bones out of the sand, sonny, if you don't watch out." "I think he'd make better dog food, sir," chimes in a black NCO accompanying the captain. Forcibly escorted to his ambulance, Dennis drives back to the funeral home, followed the entire way by two MPs.

The *Roswell Daily Record* runs the front-page headline "RAAF Captures Flying Saucer on Ranch in Roswell Region."

Early evening: At the RAAF base hospital, the horrific smell of the rapidly decomposing bodies forces the medical team to abort the preliminary autopsy.

General Ramey gives an interview on Fort Worth radio station WBAP, and reiterates the weather balloon cover story.

6:17 P.M.: An FBI memo from Dallas, Texas, to Director and SAC, Cincinnati, acknowledges that the Roswell wreckage, contrary to General Ramey's flight cancellation declaration, did indeed go to Wright Field in Dayton, Ohio, and that his weather balloon explanation had not been "borne out."

Evening: The bodies at the RAAF base hospital are sent not to Hangar P-3 for storage, but to a tent placed at the far south end of the base because of the smell. Native American enlistees from Colorado are brought in to stand guard around the tent overnight.

July 9, 1947

Early morning: The *Roswell Morning Dispatch* features General Ramey's weather balloon cover story. Most papers in the country carry both versions of the story simultaneously, though Ramey's version is more prominently featured.

Cleanup at the crash sites continues. Additional help is brought in from Fort Bliss in El Paso, Texas, to help "vacuum" the sites.

Mid-morning: Wreckage from the two crash sites is placed inside wooden crates in Hangar P-3 (one large crate, 4 x 6 x 12 feet, will hold a very special cargo), then loaded onto C-54 planes. At least one flight goes to Los Alamos, New Mexico, via Kirtland Field in Albuquerque, another to a base in Florida, but most head for Wright Field in Dayton. Sergeant Robert E. Smith of the First Air Transport Unit, who helped load the crates onto the planes, is shown a piece of the "memory metal" by another NCO, who then puts it back into his pocket.

Late morning–noon: An aircraft carrying a Secret Service envoy representing President Harry S. Truman arrives at the RAAF from Washington. Aviation pioneer Charles A. Lindbergh is also on the flight.

Transferred from the tent to Hangar P-3, the dead bodies are sealed in lead-lined body bags (due to the smell from decomposition) and loaded into the large wooden crate, then moved from the hangar to Bomb Pit #1.

Early afternoon: Officers from the RAAF base visit local newspapers and radio stations to retrieve all copies of First Lieutenant Walter Haut's original press release.

Inside the Officers' Club, Glenn Dennis meets with the nurse he ran into at the base hospital. Still sick from the day before, the nurse tells him about the aborted autopsy on the bodies, then draws a picture of what they looked like on a prescription pad. She warns Dennis that he must never discuss what she's told him or divulge her name to anyone, then leaves the club. According to Dennis, he never sees the nurse again; she was transferred to England a few days later, then allegedly killed in an airplane accident later that year.

Afternoon: The final C-54 carrying crated wreckage departs; what remains intact of the craft or escape pod itself is left behind in Hangar P-3.

Several off-duty flight crew members from the 393rd Bomb Squadron—including Robert Slusher, Lloyd Thompson, Thaddeus Love, and Joseph Osepchook—report in gear as ordered to Bomb Pit #1. The crew members see a canvas tarp covering the loading area where bombs are stored prior to loading onto an aircraft. Taxiing up the runway is a B-29, identified by the nose art as the *Straight Flush* and piloted by Captain Frederick E. Ewing of the 393rd Bomb Squadron. The waiting flight crew is ordered to turn away from the aircraft while the crate is loaded into the bomb bay, accompanied by a half-dozen armed MPs. Once secured, the crew takes off for a one-hour flight to Fort Worth. Instead of the normal 25,000-foot pressurized cruising altitude for such flights, the *Straight Flush* flies the entire way at only 8,000 feet because of MPs guarding "General Ramey's furniture" in the bomb bay. Onboard rumors of the cargo having something to do with a crashed flying saucer are confirmed to Slusher by Captain William Anderson.

The *Roswell Daily Record* runs the weather balloon cover story with the front-page headline "General Ramey Empties Roswell Saucer." A second front-page article ("Harassed Rancher Who Located 'Saucer' Sorry He Told About It"), features Mack Brazel's revised weather balloon story, which he gave to the newspaper the day before while in the security of the military. Brazel disavows the story at the end of the article, however, while the newspaper's editorial page runs an opinion piece stating that something had indeed happened, but that the army was keeping secret on the matter.

Early evening: While the *Straight Flush* taxis up the runway at Fort Worth Army Air Field, flight crew member and bombardier Lieutenant Felix Martucci recognizes an old high school friend among the greeting party of officers on the tarmac; Martucci also knows that his friend is a mortician. The crew is not permitted to disembark the aircraft. After the crate is off-loaded, Major Jesse Marcel exits the Flight Operations Building and boards the *Straight Flush* for a return flight to Roswell. As the B-29 turns around and takes off, Lieutenant Martucci is heard to exclaim, "Boys, we just made the history books!"

Evening: The *Straight Flush* arrives back at the RAAF. With the press having swallowed the weather balloon story, the "Roswell Incident" is now considered a dead issue. Marcel heads home.

The bodies are examined and kept overnight in a secure building at Fort Worth AAF.

July 10, 1947

Morning: A plane flies from Wright Field to Fort Worth AAF, then returns to Wright Field with a large metallic container.

Cleanup continues at both crash sites.

Believing he has been made a "fall guy," Major Marcel confronts CIC Captain Cavitt and demands to see his report to higher authorities concerning the events of the past few days. Marcel is refused.

Mack Brazel continues to be detained and interrogated by the military at the RAAF base.

Realizing their oversight in all of the goings-on of the past few days, a security detail from the RAAF confiscates the second box of debris from Sheriff Wilcox's office. Wilcox and his family are exhorted and threatened not to discuss what they know. Wilcox and a deputy later visit Glenn Dennis's father to warn him that his son will be in a lot of trouble if he doesn't forget about what he saw at the base.

KGFL station owner Walt Whitmore Sr., now cooperating with the military in order to save his radio license, is told to drive himself, announcer Frank Joyce, and an unidentified "sinister-looking man in a strange uniform" north of Roswell, to a small town named Lon (no longer on the map). Stopping outside a small shack, Joyce is told to wait inside. Minutes later, Mack Brazel appears and appeals to Joyce, "You're not going to say anything about what we talked about on the phone or at the radio station, are you?" Realizing his situation, Joyce responds in the negative, to which Brazel replies, "Well, good, then. Nice knowing you." Joyce responds in kind, though just before he leaves, Brazel comments, "You know, our lives will never be the same again." (As an aside, Joyce said that Brazel would call him once a year for many years thereafter, just to check up on him.)

July 11, 1947

The cleanup at both crash sites is completed. MPs and other personnel involved in the retrieval cleanup are debriefed and ordered never to discuss the incident.

British RAF officer Hughie Green drives cross-country from California to Philadelphia, Pennsylvania, listening to updates about the crash on his car radio. By the time he arrives in Philadelphia, however, there are no further updates. The story is dead, and he can obtain no further information about it.

July 12–13, 1947

Bill Brazel Jr. travels from his home in Albuquerque to his father's Corona ranch to find out what is going on. Instead, he discovers that his father is not there, nor is there any sign of the military.

July 15, 1947

After a full physical examination, Mack Brazel's interrogation and detention at the base is concluded, and he is finally returned to his ranch. For the rest of his life, Brazel remains bitter regarding his treatment at the hands of the U.S. military, just for doing what he perceived as his patriotic duty.

FBI director J. Edgar Hoover, responding to a suggestion that the FBI stay out of alleged crashed flying disc investigations, writes that he would in fact have become involved if it were not that the army in the Roswell case "grabbed it and wouldn't let us have it even for a cursory examination."

August 1947

After a rainfall near the debris field, Mack Brazel and a young ranch hand, Tommy Tyree, spot a piece of debris in a sinkhole. They leave it there and move on.

September 1947

Dr. Lincoln LaPaz, a meteorite expert secretly charged with reconstructing the object's speed and trajectory when it crashed, arrives at the

RAAF. Accompanied by CIC Master Sergeant Rickett, LaPaz spends the month traversing the entire state of New Mexico, plus parts of Arizona and Texas, interviewing ranchers and examining physical traces left by the object. It's LaPaz's belief that the object had been in trouble and touched down for repairs several times, leaving black spots that turned to glass, before exploding over the J. B. Foster Ranch. He and Rickett also find a section of pine trees that had been taken out by the object as it descended. LaPaz writes a report of his findings and sends it through channels.

October 1947

A specially modified B-29 flies what's left of the intact saucer or escape pod from the RAAF base to Wright Field.

1948

While excavating in the field, Dr. LaPaz tells Boyd Wetlauffer, a University of New Mexico archeology graduate student, about the previous year's events in Roswell. LaPaz also informs CIC Master Sergeant Rickett that he is convinced the Roswell wreckage was an unmanned probe from another planet.

1949

Flying from Albuquerque to San Antonio, Texas, General Arthur Exon and several other officers are able to simultaneously view both the Foster Ranch debris field and the crash site fifteen to twenty miles away. The vehicle tracks at each site, plus a gouge at the debris field site, remain visible.

Bill Brazel Jr., having found various scraps of debris for the past two years and kept them in a cigar box, mentions that fact in Corona. The next day, a Captain Armstrong and three others from the RAAF base turn up at his ranch and confiscate the cigar box containing the debris.

1950

Grady L. "Barney" Barnett, a Socorro, New Mexico soil conservation engineer, tells close family and friends about finding a crashed flying saucer with dead alien bodies in the plains of San Agustin, west of Socorro, a few years before. Corroboration for Barnett's story is lacking beyond those to whom he confided.

1961

Former Roswell sheriff George Wilcox passes away and is buried in Roswell. He never ran for sheriff again after the Roswell events. His wife later has a small piece regarding the 1947 events published in a local magazine.

1963

Mack Brazel, who kept much to himself following his ordeal with the military, passes away at the age of sixty-four. He is buried in Tularosa.

General Roger M. Ramey, former commander of the Eighth Air Force, passes away. He is buried in the family plot in Denton, Texas.

1965

Four-star Lieutenant General William Blanchard, deputy chief of staff of the air force and former RAAF base commander, dies of a heart attack at his desk in the Pentagon. He is buried at the Air Force Academy.

1967

"Barney" Barnett passes away and is buried in Dalhart, Texas.

1972

UFO researcher and lecturer Stanton Friedman interviews Lydia Sleppy, the woman who was prevented from placing Johnny McBoyle's Roswell crash story on the AP wire.

1978

Leonard Stringfield, an Ohio businessman and UFO researcher, interviews Jesse Marcel Sr. He subsequently self-publishes "Retrievals of the Third Kind," a limited-circulation monograph of UFO crash information accumulated over the years.

After a lecture in Baton Rouge, Louisiana, Stanton Friedman learns of a Jesse Marcel, who once held wreckage of a flying saucer in his hands, living in Houma. Friedman also hooks up with a Minnesota schoolteacher, William Moore; they begin research on the story, which yields a book two years later.

"Pappy" Henderson confides to close friend John Kromschroeder that he flew crashed UFO wreckage to Wright Field back in the 1940s.

1979

William Moore and Stanton Friedman continue to interview witnesses for their book about the Roswell Incident. Friedman also interviews Jesse Marcel for his UFO video documentary *UFOs Are Real.*

1980

The *In Search Of* television series hosted by Leonard Nimoy interviews Jesse Marcel at the Foster Ranch debris-field site. Marcel reiterates that he is certain the wreckage was nothing from this earth.

William Moore and language expert Charles Berlitz publish *The Roswell Incident,* which is based upon interviews with approximately sixty people.

1982

"Pappy" Henderson tells his wife, Sapho, about his role in the Roswell incident after seeing a story about it in a supermarket tabloid.

HBO interviews Bill Brazel Jr. at the Foster Ranch debris field for their UFO special *UFOs: What's Happening?*

1986

Jesse Marcel Sr. passes away. He is buried in Houma, Louisiana.

1987

News of the MJ-12 documents is made public to the UFO community. They are controversial documents that allegedly tell the incredible tale of the July 1947 recovery of a flying saucer near Roswell. Many believe the documents are part of a hoax by William Moore.

1988

An air force pickup truck is seen on the Foster Ranch. The driver asks ranch hand Jim Parker if the Roswell crash site is nearby.

Author Kevin Randle and Don Schmitt, a UFO investigator for the J. Allen Hynek Center for UFO Studies, reopen the investigation of the Roswell case. They are intent on proving that a weather balloon was responsible.

1989

The Center for UFO Studies conducts a preliminary archeological expedition around the Foster Ranch debris field.

The Robert Stack–hosted TV series *Unsolved Mysteries* airs an opening segment devoted to the 1947 Roswell UFO crash. The show receives wide attention, and many new witnesses come forth to aid the ongoing investigation. One witness, Gerald Anderson, claims to have been at the plains of San Agustin crash site in 1947 as a five-year-old boy. After three years of intense investigation of his claims, however, he is proven to be a hoaxer.

1990

Kevin Randle and Don Schmitt continue to interview witnesses regarding Roswell; they no longer believe a weather balloon was the cause of the event.

1991

The Fund for UFO Research brings a number of Roswell witnesses to Washington D.C., to have their testimonies videotaped and memorialized for posterity.

Joining Kevin Randle and Don Schmitt is Thomas Carey, a Philadelphia businessman and UFO investigator. Together, they try locating the archeologists who found the crashed UFO in 1947.

Former CIC Master Sergeant Lewis "Bill" Rickett passes away in Florida.

Randle and Schmitt publish their first book about Roswell, *UFO Crash at Roswell,* which is based on interviews with over 150 people.

Several anti-UFO Roswell investigators discover Project Mogul, a failed secret military project in 1947 and 1948 that tried detecting the then-impending Russian detonation of an atomic device by means of high-altitude, balloon-borne acoustic sensors.

1992

Viacom releases *UFO Secret: The Roswell Crash,* the first commercial video on the Roswell incident. Another video, *Recollections of Roswell,* is produced for the UFO community, and features the videotaped interviews of the witnesses FUFOR brought to Washington, D.C., in 1991.

Stanton Friedman and aviation writer Don Berliner publish *Crash at Corona.*

1993

The Government Accounting Office (GAO), under the urging of New Mexico congressman Stephen Schiff, announces its plans to investigate the paper trail of the 1947 Roswell events.

1994

Kevin Randle and Don Schmitt publish their second Roswell book, *The Truth about the UFO Crash at Roswell,* which is based on interviews with over 300 people.

The CBS-TV prime-time newsmagazine *48 Hours* does a pro-piece about the Roswell case.

The Showtime original movie *Roswell* premieres. The film is later nominated for a Golden Globe Award.

The air force releases its own preliminary report, put together as a result of the GAO directive. It concludes that the 1947 Roswell Incident was the result of a crashed Project Mogul balloon array. The *New York Times* uncritically accepts the air force's explanation in a front-page story.

1995

The GAO releases its report, despite being unable to find a paper trail associated with the events in Roswell; all messages and documents from the Roswell army air force base regarding the period in question had been destroyed, without apparent reason or authority.

The air force releases its second report on the Roswell case, *Fact versus Fiction in the New Mexico Desert,* which reiterates the Project Mogul theme.

1997

The fiftieth anniversary of the Roswell Incident makes the town a household name; a spate of books, TV specials, and newspaper and magazine articles culminate with a *Time* cover story in the magazine's July issue.

The air force releases its third and final report on Roswell. Entitled *The Roswell Report: Case Closed,* the document claims that the reports of alien bodies recovered as a result of the claimed 1947 UFO crash resulted from the air force's high-altitude parachute testing conducted with mannequins in the 1950s. The air force also claims that a mental process called "time-compression" is the cause of people remembering the dates wrong.

July 4: MSNBC-TV provides all-day coverage of the events taking place in Roswell.

The History Channel airs an anti-Roswell documentary that promotes the Project Mogul theme.

1998

While public interest in the Roswell incident subsides, Tom Carey and Don Schmitt team up to reinvestigate the case. Over the next four years, their investigation results in new witnesses, new information from old witnesses, and a new crash scenario.

1999

Former Roswell CIC Captain Sheridan Cavitt passes away, taking whatever he knows about the Roswell Incident to his grave. Attending medical personnel later interviewed tell of Cavitt's refusal to take pain medication during his final days, as if he were afraid of saying something he did not want to divulge. During visits from his family, he would just stare out his window and say nothing to them.

2000

Work on the "Ramey memo" begins bearing fruit. The application of computer photographic software to the 1947 photo taken by J. Bond Johnson—in which the general is seen clutching a telex in his hand—enables several teams of investigators to "read" parts of the telex.

2001

Anti-UFO Roswell investigator Karl T. Pflock publishes *Roswell: Inconvenient Facts and the Will to Believe,* which reiterates the Project Mogul theme.

2002

The History Channel does its second anti-UFO Roswell documentary, again promoting the Project Mogul theme.

The SCI FI Channel airs its groundbreaking documentary on the first archeological excavation of the debris field, sponsored by the channel and supervised by the University of New Mexico. The special (entitled *The Roswell Crash: Startling New Evidence*) garners the channel's highest-ever rating for an original special.

October 2003

The SCI FI Channel airs its long-awaited follow-up to the 2002 documentary, including an analysis of all the Historical Materials of Uncertain Origin (HMUOs) excavated in 2002.

2004

New York–based publisher Simon & Schuster releases *SCI FI Declassified: The Roswell Dig Diaries,* a complete chronicle of the historical excavation.

About the Authors

Thomas Carey, a UFO investigator with a master's in anthropology, is a former board member of the J. Allen Hynek Center for UFO Studies (CUFOS), and has been the MUFON (Mutual UFO Network) state section director for southeastern Pennsylvania since 1986, investigating local UFO sightings in the Delaware Valley. Don Schmitt, CUFOS's former director of special investigations, is the coauthor of the best-selling books *UFO Crash at Roswell* (Avon, 1991), which was the basis of the Golden Globe–nominated telefilm *Roswell* in 1994, and *The Truth about the UFO Crash at Roswell* (William Morrow & Co., 1994). He is also an adviser to the board of directors of the International UFO Museum & Research Center in Roswell, New Mexico. Carey and Schmitt are the coauthors of *The Roswell Report,* and served as technical advisers during the SCI FI Channel's sponsored excavation of the J. B. Foster Ranch debris field in September 2002.

SCI FI DECLASSIFIED

The Roswell Dig Diaries

1

THE PLAN

in·i·tia·tive noun: 1. the ability to act and make decisions without the help or advice of other people; 2. the first step in a process that, once taken, determines subsequent events; 3. a plan or strategy designed to deal with a particular problem; 4. a favorable position that allows somebody to take preemptive action or control events; 5. the right to bring a new law or measure before a legislative body; 6. a process valid in many U.S. states and in Switzerland that allows citizens to propose legislation by petition.
—*Encarta World English Dictionary,* 1999

It started with initiative. Initiative from a television miniseries about alien abductions and government conspiracies that became a cable network's impetus for publicly advocating the Freedom of Information Act. And from this came an offshoot goal: to discover the truth about what really happened outside of Roswell, New Mexico one evening in July, 1947.

A funded excavation at the former J. B. Foster Ranch in Corona wasn't on the SCI FI Channel's itinerary in April 2002, during an internal meeting in its New York City offices. The meeting was supposed to be a think-tank session for promoting *Taken,* SCI FI's then-upcoming original miniseries presented by Steven Spielberg. But it was a key plot element in the miniseries—the "Roswell Incident"—that caught the attention of Dave Howe, SCI FI's senior vice president of marketing, and Larry Landsman, SCI FI's director of special projects. Landsman was already intimately familiar with those now-famous events from 1947; while at Showtime as its director of public relations operations and photographic services, he had worked closely with producer Paul David on the network's 1994 film *Roswell,* which was adapted from Kevin Randle and Don Schmitt's book *UFO Crash at Roswell.*

Seeing *Taken* as an opportunity for SCI FI to launch its long-explored programming crusade to distinguish science fiction from science fact, Landsman helped spearhead a public advocacy and research effort by the channel. On October 22, 2002, SCI FI, allied with the Washington, D.C., lobbying firm of PodestaMattoon, traveled to the nation's capital and campaigned in support of a Freedom of Information Act (FOIA) initiative. From the National Press Club, John Podesta, former chief of staff to U.S. president Bill Clinton, would further iterate the channel's call for Congress to declassify U.S. government records on unidentified aerial phenomena.

....

For Immediate Release

SCI FI CHANNEL CHALLENGES
GOVERNMENT SECRECY
Supports New FOIA Effort on UFO Records and
Release of Report on Failure to
Conduct Scientific Research

Washington, D.C., October 22, 2002—Announcing its support for a new public effort to gain release of secret government records on unidentified aerial phenomena, commonly referred to as UFOs, SCI FI Channel today joined with John Podesta, President Clinton's former chief of staff, to call for more declassification of government records.

As part of its advocacy effort, SCI FI is backing a Freedom of Information Act initiative to obtain government records on cases involving retrieval of objects of unknown origin by the secret Air Force operations Project Moon Dust and Operation Blue Fly. Assisting SCI FI in its FOIA effort is the Washington, D.C. law firm of Lobel, Novins and Lamont.

Acknowledging the controversy that has swirled around the UFO issue for over 50 years, Podesta said, "It is time for the government to declassify records that are more than 25 years old and to provide scientists with data that will assist in determining the real nature of this phenomenon."

Explaining SCI FI Channel's involvement, Bonnie Hammer, the network's president, said, "For the past decade, SCI FI programming has explored the often-blurred line that separates science fiction from science fact. But when credible scientists conclude that 5% to 10% of UFOs cannot be explained by natural or artificial causes, we think it is worth taking a much closer look at what is clearly a real and ubiquitous phenomenon."

To bolster its point, SCI FI commissioned a report by independent journalist Leslie Kean to document the government's failure to carry out systematic scientific research into the UFO phenomenon. Kean, whose articles have appeared in major U.S. newspapers, said, "The fact is that scientists who have spent time studying, classifying, and analyzing these phenomena agree that they are real, and that it will require a sustained research effort to determine their cause."

"The public has a right to know what is being observed in our airspace, once it has been determined it is not a foreign or domestic aircraft," said Kean.

In view of a recent Roper Poll commissioned by SCI FI Channel showing that 56% of Americans believe that UFOs are real and that 72% of the public believe that the government isn't telling everything it knows about UFO activity, SCI FI also announced the formation of the Coalition for Freedom of Information (CFi). Ed Rothschild, CFi's director, said, "We have constructed a new website www.freedomofinfo.org to generate public support for greater disclosure of government records and for more scientific investigation. We are providing the public with an opportunity to get directly involved in this issue. The site contains some of the best information available on this issue as well as petitions the public can sign."

Finally, Hammer announced that on November 8, SCI FI will sponsor a symposium exploring the potential for interstellar travel and the evidence of unidentified aerial phenomena at George Washington University. Panelists will include eminent American scientists and aviation experts.

························

John Podesta,
President Bill Clinton's former Chief of Staff
Speech at a press conference
sponsored by the SCI FI Channel
Oct. 22, 2002, at the National Press Club

I thought I might start with the famous line by Admiral Stockdale: "Who am I; what am I doing here?" Or maybe I should start with who I am not: I've never been followed by a tractor beam. I've never been bathed in the glow of a white light coming from the sky. I certainly have never been taken, and while my obsession—which, for some of you, was well known while I was in the White House—with *The X-Files* earned me the title of 'First Fan,' I think I always understood the difference between fact and fiction. So, I guess you could call me a skeptic.

But I'm skeptical about many things, including the notion that government always knows best, and that the people can't be trusted with the truth. That's why I've dedicated three decades of my life, both in private practice on the Senate Judiciary Committee and in my work at the White House, to the fundamental principle of protecting openness in government. As Harold Cross, the father of the Freedom of Information Act and a well-known journalism professor at the University of Missouri, said, "The right to speak and the right to print, without the right to know, are pretty empty."

I believe that the notion of open government—the fundamental tenets of the Freedom of Information Act—are really part and parcel of our First Amendment rights. And I think it's worth going back and reminding you just exactly what those tenets really are that form the basis of that Act: that disclosure is the general rule, not the exception; that all individuals have equal right of access; that the burden is on the government to justify the withholding of a document, not on the person requesting it; and that individuals improperly denied access to documents have the right to seek injunctive relief in the courts. That's why I'm here today to add my voice to Bonnie, Leslie, and Lee's. I think it's time to open the books on questions that have remained in the dark; on the question of government investigations of UFOs. It's time to find out what the truth really is that's out there. We ought to do it because it's right. We ought to do it because the American people, quite frankly, can handle the truth. And we ought to do it because it's the law.

Let me explain what I mean by that: in 1995, President Clinton signed Executive Order 12958, which set tough standards for classifying documents, and led to an unprecedented effort to declassify millions of pages from our nation's diplomatic and national security history. Before President Clinton signed that executive order, a tiny minority of classified documents—only five percent—had a fixed classification date. Since the signing of that order, more than 50 percent of those documents are now marked for declassification in ten years or less. But even more significantly, during the five years that the executive order was in place, its policies resulted in the declassification of over 800 million pages of historically valuable records, with the prospect of many hundreds of millions more pages to be declassified in the next few years. To give you a bit of a comparison: in the previous 15 years, the government had declassified a total of 188 million pages. So I think that was a singular accomplishment of the Clinton Administration. For future generations, our history books will rely on the information contained in those declassified documents. Scholars, historians, journalists, everyday researchers around the world—not just in the United States—will explore the past and help guide our future.

But the work is not done. That order requires the automatic declassification of records that are 25 years or older, subject to a very narrow set of exceptions. And I want to note two of those: one is that the automatic declassification rule doesn't apply if it reveals the identity of a confidential human source—and I underline the word human. And I think that we're not talking about that in the cases that ought to be looked at, reviewed and declassified. The other is that they would seriously or demonstrably undermine ongoing diplomatic activities of the United States. And unless we have ongoing diplomatic activities with people who are extraterrestrials that I am unaware of, I think that exception doesn't apply, either. So, there are these cases in which documents haven't been made available to the American public. The American public is quite skeptical about the fact that the government won't make them available for public inspection. These are records that are more than 25 years old. They ought to be declassified. They ought to be released. And we ought to be able to see for ourselves what's included in them.

This morning, Dana Millbank's story in the *Washington Post* notes that our government does not always reveal the truth, the whole truth and nothing but the truth, and that even the highest leaders of our government don't always tell the facts just as they are. That's why freedom of information is so important, and the information that is included in the requests that we are discussing is so critical to be put in front of the American public, so that they can make their own judgments about the conduct of the program of investigation, as well as the facts there, that the Air Force and others have discovered.

....

Planning out the advocacy initiative did much more than provide a promotional push for a television miniseries, however; it also encouraged Landsman to renew contact with Don Schmitt, who had served as a technical advisor on Showtime's production of *Roswell.* In late spring of 2002, he asked Schmitt, now partnered with fellow UFO investigator Tom Carey, to investigate the Roswell case for the SCI FI Channel. The duo expressed that the most effective investigation would be an official excavation of the Foster Ranch debris field, in New Mexico's Lincoln County; they had proposed the idea and worked out preliminary budgets with the University of New Mexico's Office of Contract Archeology in 1999, but were unable to secure enough financial backing. Landsman took the proposal to Senior management, setting the wheels in motion with an e-mail that outlined not only Schmitt and Carey's objectives, but SCI FI's as well.

....

From: Landsman, Larry
Sent: Tuesday, May 28, 2002 1:59 PM
To: Schmitt, Don
Subject: NEW ROSWELL DIG

Don,

I left you a voicemail today as well. Keep this under your hat. I had a conversation here after our last chat. I am THRILLED to let you know that we're VERY interested to discuss partnering with you (& Univ. of

New Mexico, of course) on possible new dig at either
debris field or impact site for this September.

Let's chat!!
Larry

....

From: Landsman, Larry
Sent: Friday, May 31, 2002 5:52 PM
To: Stotsky, Adam; Guerin, Jean; Engler, Craig;
Howe, Dave
Subject: ROSWELL DIG

PROJECT STARLIGHT
NEWS ALERT

Highly Confidential

I finally connected again with research investigator Don
Schmitt today, and we spent over an hour discussing the
enormous potential for a scientific "dig" of the 1947
Roswell debris field and impact gouge (and documenting
it for a television special—either live or to be held
for a later date). Don and his longtime partner, Tom
Carey, an archeologist, have proprietary access to the
site. They are thrilled at the prospect of partnering
with us and would grant SCI FI exclusive television
rights.

Never before has a "dig" such as this been attempted
(contrary to popular belief). The only prior activity
at the site took place in 1989, when a cursory attempt
was made to create a systematic grid of the square
mile area. Funding never materialized for the
follow-up dig.

Why dig and why now?

1.) The scientific probability that predators could have
burrowed underground with pieces of the crash debris
over the course of the two days in 1947 while it lay
exposed to the elements.

2.) Advances in subterranean radar.

```
Cost?

A preliminary budget created last year by the
University of New Mexico's archeology department came
in at approximately $20,000. Don pointed out that
partial funding could come from the International UFO
Museum at Roswell.

I think if this is handled properly, and the
information about the dig (and subsequent roll-out) is
carefully controlled, this could generate enormous
attention.

Thanks!
Larry
```

....

With SCI FI on board, Landsman accompanied Carey and Schmitt to Roswell in late June, then met with UNM's OCA director Richard Chapman and William Doleman, the principal investigator assigned to head the project. It would be the first of numerous discussions regarding budgetary revisions to accommodate additional time, services, and manpower—including the use of volunteer archeologists personally selected by Schmitt and Carey—until a final contract was submitted and agreed to by all parties in late August.

During the initial meeting, Schmitt and Carey also raised the possibility of employing ground-penetrating radar (GPR), a prospecting technique used to search for buried furrows under the soil. Chapman was in full agreement and referred them to David Hyndman, the owner and principal geophysicist of Sunbelt Geophysics in Albuquerque, New Mexico. As an expert who uses noninvasive procedures to study the physics and physical processes of the earth, Hyndman would first suggest an electromagnetic conductivity (EMC) survey to measure differences in electrical conductivity within soils. After further discussion with Larry Landsman about working on the project, he'd also recommend a high-resolution metal detection (MD) survey for possible crash debris buried or washed down into the ground, plus a comparative analysis of pre- and postcrash aerial photographs of the area.

....

From: Carey, Tom
Sent: Tuesday, July 02, 2002 10:22 PM
To: Chapman, Dick
Subject: Foster Ranch Dig [Lincoln County, NM]

Dick:

Do you recommend the EMC device over the GPR for our
purposes? If yes, how soon after the procedure would
the analysis report be forthcoming? Seems to me that
there would be some raw data generated immediately
as the scan is being performed. Could you use that,
or would you wait for the formal report? Upon your
recommendation, we will lock in the EMC/GPR under a
separate contract.

BTW, Don will be in NM this weekend and early next week.
Could you benefit from a trip to the site in question?
Let me know, and I will see if Don can do it. I forget,
are we supposed to contact the BLM in Roswell, or were
you going to do that?

Tom

....

From: Doleman, Bill
Sent: Wednesday, July 03, 2002 7:59 PM
To: Carey, Tom
Cc: Chapman, Dick; Myself
Subject: Re: Foster Ranch Dig [Lincoln County, NM]

Dear Tom,

Based on information from David Hyndman of Sunbelt
Geophysics, we definitely agree that the EMC survey,
rather than the GPR, is the way to go. I believe Dick
forwarded you a copy of Hyndman's recommendations to
you. If not, please let me know and I'll get you a copy
ASAP.

Hyndman has told us that he would use a two-stage
approach. In the first stage, an EMC survey of the
entire 0.75-mile by 300-foot wide corridor would be
performed at 10-meter intervals aligned with a grid

system set up at right angles to the search area long axis by OCA personnel. The second survey would be performed at two-meter intervals on selected, smaller target areas chosen on the basis of the first survey's results.

Hyndman has told us that they can do in-field data processing and that the results should be available in 2-3 days. Hence, there would be no wait for a formal report. The choice of locations for investigatory excavations would be based on these results, with the number and density of test units being constrained by both the labor force available and the number of "anomalies" to be investigated. Given the above consideration, excavations could conceivably commence within a day or two of the end of the EMC survey.

Prior to the EMC survey and subsequent excavations, OCA would spend 1-2 days in the field to accomplish two tasks:

1) Consult with your group to locate the limits of the larger search area.

2) Use surveying equipment, rebar stakes and pinflags to establish the 10-meter EMC transect system.

As the EMC survey progresses, it is anticipated that "anomalies" would be identified and the OCA crew could follow the EMC crew to set up transect for the two-meter interval surveys of selected anomalies. This approach would hopefully provide greater efficiency and limit the delay between the identification of anomalies and their archeological investigation.

Unfortunately, I will not be available to make a field visit this weekend (July 6-7); however, both Dick Chapman and I should be available the following weekend (July 13-14) if Don is still in New Mexico.

I hope I've provided helpful answers to your questions. If not, please feel free to contact me or Dick. We both

look forward to working with you on this very
interesting project.

Sincerely,
Bill Doleman
Project Director
==========
William Doleman, Ph.D.
Principal Investigator
University of New Mexico
Office of Contract Archeology
1717 Lomas NE
Albuquerque, NM 87131
==========

....

From: Carey, Tom
Date: Tuesday, July 09, 2002
To: Doleman, Bill
Subject: Remote sensing potential for Foster Ranch

Bill,

Take a look at this website and tell me if you think
there is anything there that might be useful for our
survey at the Foster Ranch site in Lincoln County.
Thanks.

Tom Carey

(www.itres.com)
(www.itres.com/docs/casi2.html)
(www.itres.com/docs/posav.html)
(www.itres.com/docs/calibration.html)
(www.itres.com/docs/ssoptions.html)
(www.itres.com/docs/capplications.html)
(www.itres.com/docs/iqwater.html)
(www.itres.com/docs/casiswir.html)
(www.itres.com/docs/casilidar.html)

....

From: Doleman, Bill
Sent: Wednesday, July 10, 2002 4:45 PM
To: Carey, Tom
Cc: Chapman, Dick; Myself
Subject: Re: Remote sensing potential for Foster Ranch

Dear Tom,

Although I received two days of introductory training in remote sensing at the National Park Service's Chaco Center Remote Sensing Division here at UNM in 1977, I am no expert in remote sensing. That caveat aside, I have used aerial photographs extensively in past research and am generally familiar with the goals and methods of remote sensing. Nonetheless, you might want to run this question past Hyndman as well.

The Itres CASI-II airborne multispectral scanner, together with equipment and software for both geocorrection (i.e. integration into a GIS-based mapping system), and laboratory calibration, makes a nice package for acquiring and interpreting high-resolution remote sensing data in the visible light and near infrared spectrum. I'm quite impressed. The thermal and shortwave infrared systems currently in development should markedly expand the capabilities of this instrument. I believe, however, that such applications would be of limited utility for your project, for several reasons.

First, and most important, the phenomenon of interest is subsurface, while remote sensing instruments such as the CASI-II detect variations in the reflectivity of surface phenomena at various wavelengths. Although it is common for some subsurface anomalies to be reflected in surface characteristics, such as vegetation patterns, use of remote sensing would be overkill in detecting something that could just as easily be detected with aerial photography of the right scale. That's why I asked about aerial photos in our meeting.

Second, most remote sensing applications—particularly multispectral ones—rely heavily on calibration and ground truthing. Simply put, the ability to map phenomena of a certain type (pine-tree infestations, water quality) depends on the ability of the researchers to use on-the-ground observations to develop a multispectral "signature" for the phenomenon of interest. If we could develop a "signature" for our target, we wouldn't need the remote sensing! I suppose a shotgun approach might turn up anomalies of interest, but it would be an expensive experiment. Also, I suspect, if anything would be likely to detect anomalies, it would be infrared applications, which—as far as I can tell—are still in the R&D stage. Available satellite imagery (of which there is oodles) would also lack the required resolution.

Finally, the maximum spatial resolution listed (0.5 m) might not be sufficient for your purposes, whereas the EMC study would provide such resolution.

Well, that's my 2 cents' worth. I still think the EMC study is the way to go. Hope this helps. And, as always, feel free to call or e-write.

Sincerely,
Bill Doleman
==========
William Doleman, Ph.D.
Principal Investigator
University of New Mexico
Office of Contract Archeology
1717 Lomas NE
Albuquerque, NM 87131
==========

....

From: Carey, Tom
Date: Friday, July 12, 2002 12:08 AM
To: Chapman, Dick
Subject: Foster Ranch/Lincoln County Dig

Dick:

I know we talked about it at our meeting, but could you state again what the disposition would be of any

artifacts that we might find on the Foster Ranch, which
is on BLM land? Would we lose control of such an
artifact? Our preference would be that any such
artifact(s) would ultimately be displayed at the UFO
Museum in Roswell.

Thanks.
Tom

....

From: Chapman, Dick
Sent: Friday, July 12, 11:23 AM
To: Carey, Tom
Cc: Doleman, Bill
Subject: Re: Foster Ranch/Lincoln County Dig

Tom—any artifacts discovered on Federal Land are OWNED
by the Federal Agency (BLM in this case); however, the
Agencies routinely make arrangements for non-federal
museums to curate and/or display the artifacts. In this
case, we would request from BLM permission for any
objects to be curated and displayed at the Roswell
museum; it would be up to BLM to grant or deny that
permission.

Hope this helps; FYI, we still haven't received an
authorization for the project . . .

Dick

....

From: Landsman, Larry
Sent: Thursday, July 18, 2002
To: Doleman, Bill
Cc: Carey, Tom; Schmitt, Don
Subject: Foster Ranch visit & additional www link sites

Bill,

Here are some additional links to check out:

Pacificsurvey.com
They are one of the largest surveying suppliers. Look
at the Vulcan SMS, among other things.

Also, emsllc.com
The Environmental Mapping Service is a leading provider
of aerial, multi-sensor remote detection, data
collection and mapping. They offer remote sensing
support to clients and governments worldwide in
humanitarian efforts, and are committed to providing
expert solutions in GIS specializations such as UXO
(unexploded ordinance) and mine detection, subsurface
analysis and vegetation, forest research and
environmental mapping.

Finally, Epic Scan Limited provides laser scanning
modeling and could provide an exact geometric analysis
of the site. The problem is I never got a website for
them, so I don't know how to reach them. Have you ever
heard of them?

Thanks much!
Larry

....

From: Schmitt, Don
Sent: Thursday, July 18, 2002
To: Doleman, Bill
Cc: Carey, Tom; Landsman, Larry
Subject: Re: Foster Ranch visit & additional
www link sites

Hello, Bill,

Would like to hear your impressions about your visit to
the site. Also, we need to lock down some dates as soon
as possible to clear everything with the ranch owners.
How does the second or third week of September sound?

Look forward to your thoughts,
Don

....

From: Doleman, Bill
Sent: Friday, July 19, 2002 3:14 PM
To: Landsman, Larry
Cc: Carey, Tom; Schmitt, Don; Hyndman, Dave; Myself
Subject: Re: Foster Ranch visit & additional
www link sites

Dear Don, Larry and Tom:

I am Ccing Dave Hyndman so he can respond to what I'm
saying, especially if I misstate or misunderstand
anything he told me.

Sorry to take until now to report on my visit to the
Foster Ranch site. I'm pretty swamped with several
other projects, all due more or less ASAP. The life of a
contract archeologist, I guess.

Bruce Rhodes from the UFO Museum showed me around and
fed me lunch, for which I owe him big time! Very nice,
informative and helpful—and didn't ask too many
questions while trying to answer mine.

I got a pretty good look at the debris field area, and
gained some preliminary impressions which I shared with
Dave Hyndman, who I'm told is now "in the loop." He is
basing his proposal to you on what I told him. Speaking
of which, I agree with his suggestion of a renewed
metal detection survey, since he will be using
equipment with a much higher sensitivity than was
probably used before (it's the same metal detection
system that is listed on the Geonic www site he
referred us to).

I also looked at the other www sites Larry sent, and I
remain convinced that Dave's approaches are the best.
Here's why:

Pacificsurvey.com's "Vulcan 3D-I" system is for detailed
mapping of surface features, and optionally for
comparing their locations to planned or projected
locations. A data base created by such a system could
be manipulated by a GIS system to exaggerate surface
topography, and perhaps to identify "anomalies," but
I'm guessing mapping an area of 300 ft. x 0.75 mi. would

cost bogus loony dough—way more than the cost of acquiring and analyzing existing aerial photography of the area. Aerial photos, especially at the right scale and in stereo pairs (the most common format), would provide the same information value at a much lower cost. This is why Dave and I agree (I think) the expense of getting aerial photos would be well worth it. But you might want to check with Dave to confirm that.

Environmental Mapping Service's aerial remote sensing system uses the same CASI-II technology as Itres, and so is not suitable for the reasons I sent to Tom last week. EMS also offers ground penetrating radar, which Dave says will not be as effective as the conductivity survey he proposes. Briefly, the reasons for not using any of the aerial remote sensing systems are: (1) the CASI-II system only measures variations in reflectivity of the surface at various wavelengths of light, and will only inform on subsurface anomalies to the extent they are manifest on the surface. Aerial photos cost less and can also detect surface anomalies as expressed in vegetation variations. (2) If I understand correctly, you're looking for something under the surface. GPR is good for finding "hard stuff" (like bedrock) under the surface, but not soft stuff like buried "furrows," which are just dirt variations within dirt. Conductivity is much better at finding "soft" anomalies, but again I defer to Dave's greater expertise in these matters.

Finally, here are my impressions and some questions from the visit. Bruce seemed slightly vague about exactly where the debris field begins and ends. He thought the "stone marker" at the top of the hill northwest of the windmill was at the northwest end of the field, but also said he wasn't sure if it was the end or in the middle. I had him stand at the marker while I walked to the southeast for about 100 yards, and then had him tell me when I was in the middle of the field. That put me just east of the dirt track that leads to the marker from the windmill, or an orientation about 136 degrees east of true north, and on the higher part of the broad ridge the track follows. I noted a "hollow" or swale to the east of the ridge that looks a lot more like the photos you showed me.

I also asked Bruce how well known the exact location of the field and furrow is. He seemed pretty sure we were in the right place, but was also a bit vague. I ask for two reasons: (1) In order to set up a grid for the EMC survey and the test excavations, we will need to know to within a few yards, exactly where the corners are. (2) The soils on the broad ridge Bruce put me on are extremely shallow, while those to the east in the "swale" (which looks more like the photos to me, anyway) appear deeper.

Finally, I did not find any evidence of wind-blown sand deposits (that might have accumulated in the past 55 years and obscured a furrow) on the ridge. There might, however, be some washed-in deposits in the swale.

I hope this information is helpful to you.

Sincerely,
Bill
==========
William Doleman, Ph.D.
Principal Investigator
University of New Mexico
Office of Contract Archeology
1717 Lomas NE
Albuquerque, NM 87131
==========

....

From: Hyndman, David
Sent: July 30, 2002 9:48 AM
To: Landsman, Larry
Subject: Re: FW: Geophysical Surveys near Corona, New Mexico

Larry,

What are your concepts on air photo analysis? Seems to me that a study of historical air photos would be useful, looking at:

-Pre-crash
-As close to post-crash as possible
-20 to 30 years after crash, giving the vegetation time to react

I am not sure what is available without doing a search, but there is reasonable air photo coverage in New Mexico from the 1930s to the present. We often do such time-lagged analysis for engineering and environmental studies, and have found it to be a powerful tool.

Metal detection? Who knows? Our methods are a step beyond "standard" metal detection, with significantly deeper penetration and quantitative data mapping and analysis. The target objects don't need to be "metal" as we commonly perceive it, but just objects with higher electrical conductivity than soil.

I am out of the office for the remainder of the day, but hope to be in on Wednesday and Thursday.

Dave

....

From: Hyndman, David
Sent: Thursday, August 01, 2002 2:45 PM
To: Landsman, Larry
Subject: Geophysical Surveys near Corona, New Mexico, Revised Proposal

Larry,

I've attached a bid that's revised to include air photo analysis.

I am unable to make arguments for or against a metal detection survey at a UFO crash site, so I've left it in the bid as an option. For what it is worth, the Art Bell website has an account of the craft's "skin" composed of laminated bismuth/magnesium, which should be detectable.

I will be out of the office from 2 Aug. to 8 Aug. If you need to discuss the bid, please contact my associate, Mr. Sidney Brandwein.

Thanks,
Dave

....

As SCI FI's sponsored excavation proceeded apace, one concern in late June gave the channel good reason for pause: whether or not to advance-promote it. Informing the world at large would likely add up to good word-of-mouth and advertising dollars, not to mention further SCI FI's public advocacy initiative. But Landsman, Carey, and Schmitt fired off a series of e-mails that summarized the potential pitfalls of going public, and convinced SCI FI executives to rethink announcing the dig before it took place.

....

From: Landsman, Larry
Sent: Thursday, June 27, 2002 12:51 PM
To: Howe, Dave; Stotsky, Adam; Engler, Craig
Subject: ANNOUNCING DIG

Dave, Adam and Craig,

The more I thought about announcing the dig BEFORE it happens, the more I feel it is a very bad idea. I think Don and Tom (our investigators) will echo me on this—in fact, I'll solicit their opinions—but I think it would be a bad idea for a few reasons:

If we announce this is going to happen and when to the press in advance (and we would have to announce when in order for people to log on to the website):

1. The public in that area could swarm to the location (although they may not know exactly where, they could trample through nearby land), and we would have to contend with real security issues.

2. Any government faction that would not look kindly at this activity (and this is a REAL issue) may have it in its power to shut the whole thing down. Although you say it may be "fun" for the publicity, the problem is that we LOSE our 90-minute on-air special. We end up with zilch footage from the dig.

3. Besides alerting government bureaucracy, you alert all the "debunkers" who will start lining up their case AGAINST the dig and will start putting themselves out there ridiculing SCI FI.

4. Most importantly, you allow the possibility of
either a government faction or the debunkers to "plant"
evidence and taint the site.

For these reasons, I feel we should take another look
at this issue and discuss all the ramifications.

Thanks much!!
Larry

....

From: Carey, Tom
Sent: Thursday, June 27, 2002 1:20 PM
To: Landsman, Larry
Cc: Schmitt, Don
Subject: Possible Pre-Dig Announcement

Larry:

Don can give you his thoughts, but I would be surprised
if they differed much from mine. There is the potential
for bad things to happen if we announce the dig prior
to its actually taking place, or even as it's taking
place. I am against such an announcement for the
following reasons:

1. Security. There are enough crazies, wannabes, Junior
G.I. men and just plain interested who would attempt to
get out to the site to see what is going on. If this
happens, it leaves open the possibility that the owners
of the land surrounding the site will take a dim view of
this and shut us down. I don't think we can take that
chance, as this in all likelihood will be the one and
only time in our lifetimes that an undertaking such as
this will be attempted/permitted.

2. Foul Play. There are other Roswell investigators as
well as Roswell debunkers out there (who shall remain
nameless) who wish us ill and would do everything in
their power to prevent, ridicule or sabotage this dig,
if they can. If the dig turns up nothing, they will have
had a chance to prepare "See, I told you so" statements
to ridicule us further.

3. Foul Play II. With advance notice, who knows what government agency or military command might attempt to mess things up for us with a preemptive trip to the site, to destroy site integrity and render our findings as suspect? I would not put this out of the realm of possibility, crazy as it sounds.

4. Luck. I don't want to jinx our luck with a premature announcement. That has always been my experience in anything I have attempted to do.

Tom Carey

....

From: Schmitt, Don
Sent: Thursday, June 27, 2002 3:42 PM
To: Landsman, Larry
Cc: Carey, Tom
Subject: Re: FW: ANNOUNCING DIG

Larry, et al.

First, I totally concur with TC for all of the reasons he cited. Also, it can be demonstrated that within recent times (past 15 years) the military has been observed and ejected from the private ranch property for "nosing around" that very pasture of land. Second, yes, the site is highly remote, but as evidenced by a local military plane crash which took place in this general area within the past 20 years, the event became a tourist attraction for the locals. We have to remember that not much happens within 50 square miles of this region, and word does get around.

And last, let's not telegraph our actions if we don't have to. Lest we forget we're playing with the big boys here, and we not only have to respect that fact, but also anticipate it. We have a chance to create our own success here. We cannot afford to make any mistakes which may benefit the opposition.

Your thoughts?
Don

....

From: Carey, Tom
Sent: Thursday, June 27, 2002 7:37 PM
To: Landsman, Larry
Cc: Schmitt, Don
Subject: Re: FW: ANNOUNCING DIG

Larry:

O.K. The two biggest concerns are:

(1) The chance for military/government chicanery [some gov't/military entity knows the site and has periodically kept track of it for the past 55 years]. Their vehicles have been seen and chased away. They did a good job of "vacuuming" the site 55 years ago. Who knows what they could do today?

(2) The site might be remote, but this is what will happen if word gets out: first, the local ranchers and their kids from adjoining ranches in the Corona area will mosey on down to have a look. Then the townsfolk from Roswell will drive on up to see, too. Then all those within a day's drive will show up, and finally, since Roswell is the most famous UFO case of all time— a household word—people will come far and wide [like going to Mecca] just to be there. Chaos will reign, and the ranch owner, since the site is surrounded by a private, functioning ranch, will shut us down. That's my biggest fear. The site may be remote, but people are much more mobile than in 1947, and quite a few people who had neither radio, TV or four-wheels still got there back then.

Tom

....

From: Howe, Dave
Sent: Friday, June 28, 2002 1:37 PM
To: Engler, Craig
Subject: Roswell Dig

Craig:

Upon further investigation, Larry has talked to a number of key individuals at Roswell, as well as

investigators who've worked there. He's come back with
some very compelling arguments from a number of key
people for not announcing the dig or covering it on
SCIFI.com. The threat of the dig being sabotaged/pre-
empted or shut down appears to be very real. Therefore,
on this basis, I don't believe we should either cover
it or announce on SCIFI.com. I know you'll be
disappointed, Craig, but it's just not worth the risk.
I know that having the dig stopped would be newsworthy,
but not as newsworthy as actually digging something up.
We need it to go ahead.

Let's discuss at the next Starlight meeting.

D.

....

Secrecy remained a constant watchword not only before, but also throughout the dig. SCI FI would even go to the extra expense of hiring Fortress
Security and Investigations Inc. (FSI) to protect personnel at the Corona
debris field and ensure that no outsiders would tamper or "salt" the site;
that is, deposit foreign elements that weren't already in the soil, to help suggest that something had been uncovered.

Plans to film the dig, meanwhile, continued to move quickly and quietly; by August SCI FI had entered into an agreement with MPH Entertainment, a Burbank, California–based production company responsible
for films like *My Big Fat Greek Wedding* and documentaries such as *The Lost
Dinosaurs of Egypt*. The agreement called for MPH to produce a two-hour
documentary that recalled key details of the Roswell Incident and introduced new evidence to the case, including the excavation at the Foster
Ranch site. The documentary, to be hosted by *The Early Show*'s Bryant
Gumbel and directed by MPH's Melissa Peltier, was eventually called *The
Roswell Crash: Startling New Evidence*. Both MPH and UNM's Office of
Contract Archeology would require budget adjustments prior to the excavation. MPH mounted on the additional cost of shooting aerial footage
from aboard a helicopter, which Tom Carey and Don Schmitt had requested a month earlier to investigate a possible third impact site. For
OCA's Bill Doleman, the supplemental funds were to compensate for administrative, logistical, and technical factors that couldn't be anticipated
beforehand.

Unsurprisingly, legal and liability issues would also surface as the excavation grew closer. Ironclad nondisclosure and appearance releases were drafted to prevent anyone participating in the dig from disclosing details prior to airing the documentary, plus to cover the potential hazards inherent in letting untrained volunteers work in a deserted, and possibly unsafe, climate. Then there was the matter of hotel and travel accommodations— the believed crash site of the Roswell Incident was actually more than ninety minutes outside downtown Roswell, with other towns or rest stops few and far between.

By early September, all paperwork had been submitted, contracts and release forms signed, and reservations booked. Only two final, yet extremely important, tasks remained for OCA project director Bill Doleman before all parties headed to New Mexico: he had to give the inexperienced SCI FI and MPH crews, plus the volunteers, a list of what to bring for the excavation, and he had to submit an archeological test plan for the Foster Ranch project to the Bureau of Land Management (BLM), which owned anything found on the site. Doleman would submit the test plan to the Roswell BLM office on September 12, less than three days before preliminary work was scheduled to take place at the site. (A modified version of the plan, seen here, was submitted in November.)

....

From: Doleman, Bill
Sent: Thursday, September 05, 2002 1:26 PM
To: Landsman, Larry; Schmitt, Don
Cc: Myself
Subject: Volunteers/liability issues,
driving time & motels

Larry and Don:

(1) Volunteer liability issues:
I finally got the lowdown on volunteers from UNM's Risk Management division (after getting the wrong info from UNM's legal dept.). No volunteer release forms are necessary. All I need to do is get a list of names and social security numbers to Risk Mgt. before the project starts (date of birth is OK for those wary of providing an SSN). The Office of Contract Archeology will then be billed $2 a head to get them listed on UNM's blanket volunteer insurance coverage.

Don, if you could get that info to me ASAP, I'll forward
it to Risk Mgt. By the way, I told them nothing of the
nature of the project; just that it's a volunteer
archeology "dig."

(2) Driving time and motels:
I don't know if you all have finalized plans to stay in
Roswell or Vaughn (Larry was asking the other day, I
think). Vaughn offers the least travel time, but Bill
Parks of MPH Entertainment said their crew would
probably be staying in Roswell as they have interviews
to do there. My analysis (see below and attached Excel
file) suggests a 77-minute trip from Vaughn, and 89- or
112-minute trips from Roswell, depending on the route
taken.

Once the town we're staying in has been finalized, I
planned to make motel reservations only for myself and
OCA employees, and any volunteers I might scare up here
(my wife will join us for a few days if Mr. Bogel is OK
with our two dogs—anybody got his phone number?). I
hope this is OK. I got a list of six Vaughn motels from
an internet site where almost all towns and their
motels are listed (click on New Mexico, then various
cities/towns at http://www.usa-lodging.com/). There
are 21 listed for Roswell, a much bigger town.

So, what do you think?

Cheers,
Bill

Mileages & estimated driving times to crash site from Vaughn and Roswell (assumes 60 m.p.h. on paved roads, 40 m.p.h. on gravel county roads; mileages based on DeLorme Atlas [1 inch=4.7 mi.]):

		Inches	Miles	Time (minutes)
1	From Vaughn			
	US 54 NM 247 (paved) : Vaughn to CR E010	11.6	54.5	55
	CR E010/018 (gravel) : to site	3.2	15.0	23
		Total	69.6	77
		Inches	Miles	Time (minutes)
2	From downtown Roswell via US & NM 247			
	U.S. 285 & NM 247 (paved): Roswell to CR E010 19	19	89.3	89
	CR E010/018 (gravel): to site	3.2	15.0	23
		Total	104.3	112
		Inches	Miles	Time (minutes)
3	From downtown Roswell via NM 246 & misc. county roads			
	NM 246 (paved) : Roswell to CR F008	8.9	41.8	42
	"CR F0018, etc. (gravel) : to site"	6.7	31.5	47
		Total	73.3	89

....

From: Schmitt, Don
Sent: Thursday, September 05, 2002 4:35 PM
To: Landsman, Larry
Cc: Doleman, Bill
Subject: Re: Volunteers/liability issues,
driving time & motels

Gentlemen,

The volunteers break down as follows:

1. New York
1. Indiana
1. Arizona   willing to share 1 room
1. Missouri
1. Colorado   1 room
1. New Mexico   2 rooms
1. Texas
1. Texas
1. Texas   2 rooms
1. Texas   1 room
1. Wisconsin
1. Wisconsin   camper at site
                  Total number of rooms in Vaughn: 5

I will get the personal particulars ASAP.

Don

....

From: Doleman, Bill
Sent: Monday, September 09, 2002 3:12 PM
To: Landsman, Larry; Schmitt, Don; Carey, Tom; Parks,
Bill; Lublin, Christina; Crabb, Kim; McPherson, Kelly
Cc: Myself
Subject: Footgear and stuff

Dear all:

Tom asked about footwear for the site. Here's my
response. I've no doubt left something out, but it's a
start:

A TOP-DOWN APPROACH TO THINGS EVERYBODY SHOULD HAVE IN
THE FIELD

We'll be working in the late summer in high desert at
an elevation of over 5,800 feet. That means lots of
sunlight (like the beach), but less atmosphere to
protect us from it. Mid-September is usually dry—1.8
inches average monthly rainfall. The "desert" also
means potentially variable weather, although late
summer is the least variable. The weather through the
day can change a lot, too. The average high for
September 20 is 83.2 degrees Fahrenheit, with 95
percent chance the high will be between 65 and 99
degrees Fahrenheit, and a 67 percent (2/3) chance the
high will be between 74 and 91 degrees Fahrenheit. That
is, it will probably be warm, possibly cool or hot.
Average low is 50 degrees Fahrenheit. Wind and rain are
generally not a problem in the late summer, but we're
in the last stage of the "monsoon," which means
afternoon thunderstorms are still a possibility.
Occasionally, we'll get all-day rain as a spin-off from
a hurricane that's hit the Gulf Coast.

Although Tom asked about footwear, let's start at
the top:

Hat: Anything with a brim (like a gardener's hat) is
good. Baseball caps are minimally OK, but I wear mine
with a bandana to protect my neck and ears (don't want
to be a redneck).

UV-blocking sunglasses.

Sunscreen. Lots of it.

Hand cream, and lots of it (especially important for
digging volunteers).

Work gloves are important for digging volunteers.

Knee pads are useful for digging volunteers.

A variety of clothing for variable weather, including
long- and short-sleeve shirts, and a windbreaker; we

generally "layer" and "de-layer" as the weather changes throughout the day.

Shorts are OK, unless you're digging, which involves a fair amount of on-the-knees activity. Generally, tho', I recommend long pants, preferably jeans.

Now, to footwear: This is NOT sandal country, so leave your Tevas in the motel! The best footwear is over-the-ankle work or hiking boots. This is because, although the vegetation is generally low, there's enough cactus and other spikey stuff (yuccas) that you want your feet and ankles protected, especially if you're not used to looking for it.

We'll have some water jugs out there, but bring your own food and other drinks.

That's all I can think of for now.

Cheers,
Bill
==========
William Doleman, Ph.D.
Principal Investigator
University of New Mexico
Office of Contract Archeology
1717 Lomas NE
Albuquerque, NM 87131
==========

....

From: Doleman, Bill
Sent: Monday, September 09, 2002 3:23 PM
To: Landsman, Larry; Schmitt, Don; Carey, Tom
Cc: Parks, Bill; Lublin, Christina; Crabb, Kim; McPherson, Kelly; Myself
Subject: The master plan and schedule

Dear all:

Attached are two Word doc files for informational purposes. The first contains the testing plan I submitted to the Roswell BLM on Friday, 9/6. This was required for our excavation permit; it describes all

the major project activities and their purposes, and
provides a semi-firm schedule.

The second is a statement of OCA's qualifications,
should anyone be interested. It gives the lowdown on
our history and previous work, etc.

Hope you all find these useful as we gear up for an
exciting project. If you need either in another format,
please let me know and I'll try to accommodate you.

Cheers,
Bill
==========
William Doleman, Ph.D.
Principal Investigator
University of New Mexico
Office of Contract Archeology
1717 Lomas NE
Albuquerque, NM 87131
==========

....

Notes:

1. The following testing plan was submitted to the USDI Bureau of Land Management Roswell Field Office on Sept. 12, 2002, and approved by the Roswell field office, the state BLM office, and the State Historic Preservation Division.

2. Attachments referred to in the document are not included as (a) the OCA excavation forms are not unique to the project and (b) the exact map location will be provided in the final report after Nov. 22, 2002.

3. The addendum that follows reflects changes in methods that were approved through consultations with BLM Roswell Field Office personnel in the field.

....

ARCHEOLOGICAL TESTING AND REMOTE SENSING STUDY PLAN FOR FOSTER RANCH IMPACT SITE
submitted to
USDI Bureau of Land Management Roswell Field Office
by William H. Doleman, Ph.D.
Principal Investigator
UNM Office of Contract Archeology

INTRODUCTION

This testing plan describes field and laboratory methods for a program of remote sensing and limited archeological testing to investigate a 300-ft. by 0.75-mile area for the purposes of detecting physical evidence of a low-angle extraterrestrial vessel impact in 1947. The area does not currently have a Laboratory of Anthropology (LA) site number, but one will be acquired from the state Historic Preservation Division's Archeological Records Management System in Santa Fe, should the proposed testing activities reveal prehistoric or historic remains warranting such designation. If so, a Laboratory of Anthropology site form will be prepared. A preliminary visit to the location revealed the presence of a few, very scattered chipped stone artifacts in the area, but no artifact concentrations or features warranting site designation.

The archeological testing work will be accomplished by volunteer excavators under the supervision of a project director (William Doleman) and trained archeological crew members from the University of New Mexico Office of Contract Archeology (OCA, 1717 Lomas NE, Albuquerque, NM 87131). The remote sensing component of the investigation will be conducted by Mr. David Hyndman of Sunbelt Geophysics (P.O. Box 36404, Albuquerque, NM 87176). Mssrs. Don Schmitt and Tom Carey—recognized Roswell Incident researchers—will serve as technical advisors to the project. The project is being funded by USA Cable and the SCI FI Channel.

The area to be tested—commonly referred to as the "skip site"—lies on BLM and private lands on the Foster Ranch in Lincoln County, NM. The exact location of the target area lies [CLASSIFIED].

The project's investigative strategy entails using (1) nondestructive remote sensing techniques and (2) standard archeological subsurface testing methods to search for physical evidence of a vessel impact. Two possible kinds of evidence are anticipated. The first is evidence of the impact in the form of actual modification to the original ground surface caused by the vessel impacting at a low angle, reportedly producing a "gouge" or "furrow" in the land surface. This furrow may have been filled in through natural erosion and deposition processes during the 55 years that have passed since the impact is thought to have taken place. The second kind of possible evidence is a "debris field" that was reportedly produced by the impact, and associated partial disintegration and scattering of portions of the vessel's exterior. Much of this material was apparently removed within a few days of the impact, but some material may have been missed and been subsequently buried by natural processes, including erosion, deposition, and bioturbation.

All investigative activities will be documented in the field including establishment of a grid system tied to an existing permanent marker at the site as well as to nearby BLM benchmarks. The grid system's orientation will be determined by the suspected orientation of the impact "furrow" and debris field. Placement of archeological test units within the grid system will be based on three sources of information: (1) results of the aerial photograph inspection and identification of surface anomalies, (2) locations of subsurface anomalies as revealed by the EMC study, and (3) available anecdotal accounts of the locations of the impact and the resulting debris field.

NON-DESTRUCTIVE REMOTE SENSING

The remote sensing component of the investigation will entail three activities and will be conducted by Mr. David Hyndman of Sunbelt Geophysics.

Aerial Photograph Study

The first remote sensing activity is inspection of available aerial photographs of the project area taken both a few years before and a few years after the impact reportedly took place in 1947. The goal of this study will be detection of changes in the ground surface and/or vegetation patterns that reflect the effects of the impact. The provenience control grid system will be marked on copies of the aerial photographs as an aid in targeting changed areas for archeological investigation.

Electromagnetic Conductivity Survey

The second remote sensing activity will consist of a two-stage electromagnetic conductivity (EMC) survey of the ground. The EMC equipment consists of an EM-31 device that rides on small bicycle-like wheels across the surface in a non-destructive fashion. The EM-31 detects subtle variations in ground conductivity that in turn reflect variations in substrate consistency (compaction, moisture content, soil mineralogy, etc.). Such subsurface variation would be produced by a buried feature such as the hypothesized furrow and—if present— should be detectable by the EM-31. The first-stage EMC survey will consist of linear transects spaced at 10-meter intervals and oriented at right angles to the estimated axis of the impact and debris field. The EM-31 is capable of in-field real-time data processing and identification of subsurface anomalies. Subsequent to the 10-meter interval survey, a second-stage EMC survey of definite and possible anomalies will be conducted at 2-meter intervals oriented either across anomalies identified by the first stage survey or, as before, across the impact zone's primary axis. The results of the EMC surveys will be used to determine optimum locations for using archeological test units to investigate subsurface anomalies.

Metal Detection Survey

The third remote sensing activity will consist of a high-resolution metal detection survey of suspected debris locations selected through consultation with the project's technical advisors. The goal of this activity will be the discovery of buried vessel debris.

ARCHEOLOGICAL TESTING

As noted, archeological test excavations will be conducted for the purposes of exploring subsurface anomalies revealed by the EMC survey, as well as to search for buried extraterrestrial materials. The testing program will incorporate standard archeological methods and will include photography, mapping, in-field documentation of surface artifacts, and excavation of exploratory test pits for the purposes of determining subsurface stratigraphy and to investigate anomalous localities. The following archeological activities will be conducted:

Grid System and Provenience Control

The OCA crew will use a transit, stadia rod and metric tapes to establish a datum and a standard metric three-dimensional cartesian grid system over the site to facilitate mapping and recordation. Locations of all major control points will be recorded using averaged GPS readings ($>=15$ m accuracy) and will be marked on copies of available aerial photographs of the project area. Once the location and estimated axis of the target impact zone and debris field have been determined by consultation with the project's technical advisors, a central datum will be established and permanently marked with a steel rebar and stamped aluminum cap, which will display the site number should one be assigned. A baseline that follows the target area's central axis will then be established using the same equipment to measure locations and relative elevations. Rebars and aluminum tags displaying coordinates will be placed at 50-meter intervals, while large nails and/or pinflags will be used to mark 10- and 2-meter intervals. The baseline will provide control points for laying out the EMC survey transects at right angles as discussed above. In addition, the baseline control points will be used as subdatums to measure and map the locations of metal detection survey localities, as well as all archeological test units and any other relevant features or artifacts.

Once the datum and baseline have been established, a generalized field map of the target area, grid system, and all relevant investigative features will be prepared using the transit data gathered in the course of establishing the grid system. As EMC survey transects and archeological test units are established and investigated, they will be added to the map. In addition, data required for generating a contour map will be gathered. In the laboratory, all transit data, including descriptive information, will be entered into an OCA computer database and coordinate conversion system that will produce cartesian coordinates for use in a CAD mapping system.

Test Excavations

As noted above, the locations of individual archeological test units within the grid system will be based on three sources of information: (1) results of the aerial photograph inspection and identification of surface anomalies, (2) locations of subsurface anomalies as revealed by the EMC study, and (3) available anecdotal accounts of the locations of the impact and the resulting debris field. Test unit locations will be measured and mapped using transit and stadia rod as described above. Excavations will be conducted by 6–18 volunteer excavators under the supervision of an OCA project director and two archeological assistants. All recovered artifacts will be provenienced to a minimum of 1-meter accuracy and bagged and cataloged in the field according to standard OCA procedures. If any human-made archeological features are encountered, they will be left intact for future investigations, and the test unit will be adjusted accordingly.

Prior to excavation of any test units, however, one or more units designed to determine natural stratigraphy will be excavated and a provisional stratigraphic system will be developed for use during the test excavations. The target area lies in rolling terrain typical of karst geological environments. Karsts form in broad areas underlain by limestone and exhibit topography characterized by sinkholes and other features formed through dissolution of rock's calcium carbonate by infiltrating precipitation, which is mildly acidic owing to dissolved carbon dioxide. Surface deposits in the area appear to consist largely of in-situ weathering products, including limestone clasts and fine-grained material of a silt-loam texture, some of which may be of eolian origin.

Once the local natural stratigraphy has been determined, the locations of particular archeological test units will be chosen and mapped. Test units will be begun as 0.5 x 0.5 m pits, but may be expanded for convenience after reaching depths greater than ca. 50 cm. Excavation will be by hand and vertical control by both 5-cm levels and natural stratigraphic units (where detectable) will be maintained. All excavated fill will be screened using 1/8-inch mesh, and samples of fill from each level will be collected for possible future analysis (including microscopic inspection).

The volunteer excavators, their supervisors and the project director will keep detailed notes on test unit stratigraphy and any recovered artifacts and FS numbers using standard OCA grid excavation forms, and all units will be given a unique study unit number. In addition, plans and profiles of all units will be photographed and drawn where deemed necessary. All recovered artifacts—both Native American and extraterrestrial—will be entered into a field specimen catalog, with each unique horizontal and vertical provenience (e.g., grid coordinates, 5-cm level and natural stratum) receiving a unique field specimen (FS) number.

Where possible, locations of specific artifacts and other features of interest within individual excavation units will be measured and recorded to the nearest centimeter. Examples of standard OCA excavation forms are attached.

All recovered artifacts will be collected. All Native American artifacts encountered on the surface or in test excavations will be placed in envelopes or bags with labels showing the excavation unit's provenience information and FS catalog number, and—at the end of the project—transported to the OCA laboratory for cleaning and subsequent analysis. Disposition and analysis of any suspected extraterrestrial remains will be determined by pre-field consultation with BLM personnel and the project's sponsors. It is presumed that any such materials will be transported to an appropriate laboratory for scientific evaluation. Should any Native American features be encountered, they will be described, photographed and mapped. Following such recording, they will be covered with plastic and reburied for future study.

All backdirt produced by excavation of test units will be placed on plastic sheets next to the units, and units will be backfilled upon completion.

Provenience Control and Integrity

As noted above, locations of all test units will be recorded and mapped using the metric cartesian grid system established. Within individual units the locations of recovered remains will be measured vertically within 5-cm and natural stratigraphic levels, and horizontally with the same accuracy as the unit's location, and where possible to the nearest centimeter.

In addition, measures will be implemented to ensure "provenience integrity"—that is, to avoid as much as possible the intentional or inadvertent introduction of artifacts into excavation units, either by members of the project team or unknown parties. These measures will include the following:

1. Vehicular traffic will be limited to existing roads and two-tracks. Foot traffic will be limited to specific paths that follow the grid system axis and cross-alignments chosen to maximize access to test units. These measures are designed to limit disturbance of potential test locations prior to and during archeological investigation.
2. Each uncompleted excavation unit will be covered with plastic and backdirt at the end of each day.
3. Excavation units will be photographed on a regular basis, including:
    a. photographing the natural surface of each test unit prior to beginning excavation to ensure it has not been disturbed,
    b. photographing the grid's surface at the completion of each excavation level, and
    c. photographing the grid's surface at the end of each day prior to covering and again each morning after uncovering and prior to continued excavation (any disturbance during the intervening time should be evident).
4. Sediment samples taken from test units will be sealed with high-quality tape to ensure their integrity during and after transport.

LABORATORY PROCESSING

Artifacts and sediment samples will be transported to the laboratory at OCA. Sediment samples will be left sealed and curated in a locked cabinet, while Native American artifacts will be processed using standard OCA methods that include checking against the FS catalog, cleaning with water, and rebagging and labeling. All collected Native American artifacts will be briefly described and photographed and/or drawn if their nature warrants. As noted above, it is presumed that any suspected non-terrestrial materials will be transported to an appropriate laboratory (to be determined by consultation between BLM and the project's sponsors) for scientific evaluation.

REPORTING

Following the fieldwork and laboratory processing components of the project, a detailed report will be prepared that describes the project and its results, including field and laboratory methods, locations of EMC survey transects and archeological test excavation units, the results of all excavations (artifacts recovered, the nature of excavated deposits and any subsurface features encountered), and the results of laboratory analyses. The report will include reviews of the project area environment, prehistory and history, and will be accompanied by illustrative photographs and graphics.

Graphics will include a master site map prepared from field-collected transit data that depict the locations of all remote sensing studies and archeological test units, as well as other relevant site features. Although a separate report on the results of the remote sensing studies will be prepared by Dave Hyndman of Sunbelt Geophysics, important remote sensing study results will be included on the archeological report map.

PROJECT SCHEDULE

The fieldwork phase of the project will consist of three separate activities: (1) establishment of a cartesian grid system and control point, (2) a two-stage electromagnetic conductivity (EMC) survey of the ground within the hypothesized impact zone and debris field, and (3) excavation of archeological test units in selected locations for the purposes of (a) investigating subsurface anomalies revealed by the EMC survey, or (b) examining subsurface deposits for the presence of non-terrestrial materials.

The tentative project start date is Tuesday, September 17, and the fieldwork is slated to continue for at least eight days, through Tuesday, September 24. OCA staff will be on site on Monday, September 16, however, to consult with the technical advisors for the purposes of determining the exact location of the targeted study area. Arrangements have been made to continue the project on a daily basis, should discoveries warrant it. The planned project schedule is as follows:

Sept. 16: OCA crew meets with technical advisors to determine study area limits

Sept. 17: OCA crew begins establishing grid system, datum, baseline, etc.

Sept. 18: OCA crew completes grid system and 10-meter EMC survey begins

Sept. 19: 10-meter EMC survey continues; OCA crew consults with EMC survey crew, using analytical results to begin choosing and marking locations for 2-meter EMC survey

Sept. 20: 2-meter EMC survey completed; OCA crew consults with EMC survey crew, using analytical results to choose and mark locations for archeological testing units

Sept. 21: Volunteer excavators arrive and begin test excavations under OCA crew supervision; high-resolution metal detection survey begins

Sept. 22: test excavations continue; remote sensing activities may continue

Sept. 23: test excavations continue; remote sensing activities may continue

Sept. 24: test excavations continue; remote sensing activities may continue

Sept. 25–?: test excavations may continue; remote sensing activities may continue.

ATTACHMENTS

1. Examples of standard OCA excavation forms (grid excavation form, field specimen catalog, plan/profile form, transit mapping form)
2. Map showing possible locations of 300 ft. x 0.75 mi. study area

ADDENDUM TO ARCHEOLOGICAL TESTING AND REMOTE SENSING STUDY PLAN FOR FOSTER RANCH IMPACT SITE

Three changes were made to the field methods proposed in the original testing plan for the Foster Ranch project. These changes were approved by Roswell Field Office personnel.

1. 10-cm excavation levels were used instead of the originally proposed 5-cm levels. This modification was made in order to reduce paperwork and time required for each test unit, and occasioned by the fact that only six volunteer excavators were available, versus the 12 originally stipulated. The concomitant reduction in vertical provenience accuracy was not expected to have any substantial affect on the validity of project results.
2. The use of backhoe trenches to search for furrow evidence and to investigate anomalies revealed by the electromagnetic conductivity or metal detection surveys was implemented to increase the chances of discovery and to allow investigation of depths greater than those possible in 50x500-cm test pits.
3. All soil samples and non-natural materials recovered in the course of the test excavations were transferred to a locked security van at the end of each day. At the end of the project, the materials were transferred to a lock box at the Wells Fargo Bank in Roswell.

STATEMENT OF QUALIFICATIONS—OFFICE OF CONTRACT ARCHEOLOGY/UNIVERSITY OF NEW MEXICO
April 2002

The Office of Contract Archeology (OCA) is a cultural resources management program established within the University of New Mexico in 1973 for the purposes of aiding clients in all aspects of Section 106 compliance, and incorporating this service into a strong research program.

A division of the internationally recognized Maxwell Museum of Anthropology, OCA has taken a leading role in innovative, large-scale, multiple-task and interdisciplinary cultural resource studies requiring simultaneous field and analytical efforts throughout New Mexico and adjacent states. OCA's central location and experience with numerous large-scale projects has resulted in the firm's development of a physical plant, management strategy, staff structure, and field, laboratory, analysis, and reporting processes geared for high-quality, innovative and rapid-response action demanded by simultaneous multiple-task projects. This staff and infrastructure, together with OCA's geographically broad history of research in New Mexico, enables OCA to efficiently perform cultural resource investigations and develop scientifically-based management solutions for projects of all sizes.

Since 1973, OCA has completed over 800 cultural resources management related projects, and has established a strong track record in diverse research, educational and publication programs. Highlights of this history include:

- Archeological surveys
- Archeological excavations
- Archeological testing and data recovery plan preparation
- Research design development
- Historic Building Inventories
- HABS/HAER documentation
- Literature surveys and assessments
- Popular publications and brochures
- NRHP nominations (both historic and prehistoric)
- Ethnological/TCP studies
- Agency coordination
- NAGPRA consultations, burial excavations, and reburials

Staff

The Senior staff includes a Director and an Associate Director who serve as Principal Investigators on contracts and grants; a Computer Systems Analyst/Programmer; a CAD/GIS Specialist; four Project Directors, a Laboratory Director, four Crew Chiefs, and an Administrative Assistant. In addition, OCA has an experienced pool of numerous Assistant Archeologists well qualified for survey, excavation, and laboratory tasks, and we have long-term consultant relationships with analytical specialists for pollen, macrobotanical, and editorial tasks.

Most of our permanent staff have been with OCA for 15 years or more. Senior staff regularly update critical skills by participation in workshops, conferences, and classes, particularly those devoted to evolving cultural resource laws and regulations, specialized archeological knowledge, and computer applications.

Facilities

OCA's facilities are located at 1717 Lomas Blvd. NE on UNM's North Campus. They total 8560 sq. ft. of space housing laboratories (3100 sq. ft.), office space (4000 sq. ft.), and artifact and equipment storage (1460 sq. ft.). A secure fenced vehicle yard is attached to the building. Space used for general administration includes a reception area with a dedicated FAX line, and offices for the Director, Associate Director, and Administrative Assistant. The facility also has separate rooms for CAD/GIS, data processing, and report production. Eight fully equipped secure offices are dedicated solely to Project Directors. The facility contains analysis pods equipped to support 15 individual analysts, with desks, files, and network-linked microcomputers. OCA also houses a 5,000+ volume library, and a conference room for meetings and presentations.

Field Equipment, Vehicles, Lab Equipment, Office Equipment

OCA owns a wide range of equipment and vehicles necessary for fielding several large projects simultaneously in survey, excavation, and related operations. Included in our inventory are all of the necessary cameras, compasses, transits, alidades, wheelbarrows, screens, and the full range of hand tools and personal equipment necessary to perform large surveys and excavation projects. We also have a number of more specialized items, such as tree-ring and archeomagnetic sampling kits.

OCA owns a 4WD Suburban, two 4WD cargo trucks, and a 15-passenger van. We also have immediate access to the UNM automotive fleet for short-term rentals, as-available access to 4 x 4 vehicles maintained by the Departments of Geology and Biology, and open purchase orders for off-campus vehicle rentals as needs arise.

Laboratory facilities include sinks, ample working and storage space for large-scale artifact processing and analysis, and a full range of basic analysis equipment, such as binocular and petrographic microscopes, triple-beam and digital balances, flotation devices, digital calipers, compasses, and microphotography equipment.

OCA has all of the necessary hardware, software, and staff experience to carry out all computer-related tasks. Our extensive array of in-house software and hardware gives us total capability for a wide range of computing projects. We are well equipped for the various word processing, data management, analysis, graphic display, cartographic and report production tasks associated with cultural resource management contracts. We have tremendous flexibility to tailor our system to the requirements of each project. We have a full-time computer specialist, who is responsible for maintenance and enhancement of computer facilities at OCA. His duties include the selection and procurement of software and hardware, as well as creation of custom input and output routines. He is also involved in the design of data entry and field data recording forms, and manages our Novell v5.1 50-user network.

Data Management and Reporting

The data management structure presently in place at OCA has been designed and refined in response to several large-scale multiple task services contracts with the U.S. Army Engineer District, Ft. Worth; the New Mexico Bureau of Land Management; the U.S. Army Corps of Engineers, Albuquerque District; the New Mexico State Highway and Transportation Department, and private pipeline, powerline and optical fiber cable firms. All of these multi-year contracts have required frequent tasks with very short start-up times, often large field crew sizes, and rapid reporting turnaround times. To deal with these task requirements, OCA has developed a data management structure which emphasizes computerization of field excavation and survey data during the course of fieldwork, preliminary analysis and reporting stages. This structure has resulted in producing comprehensive data compendia which have been used by agencies to obtain compliance to proceed with project actions shortly after fieldwork is completed. Our data management structure also speeds final report production schedules by emphasizing integrated use of survey and excavation databases for analysis, mapping, and graphics.

Research Capability

OCA houses and staffs its own research library, which contains extensive collections of unpublished and limited-distribution cultural resource management generated data for the entire state of New Mexico, the Four Corners region, and West Texas. Complementing this archive is a map file of 7.5' and 15' USGS quadrangle sheets for the same geographic regions. On-line searches of the New Mexico Historic Preservation Division NMCRIS statewide archeological records are made via computer and as necessary by visiting the NMCRIS paper archives. Other library and archival resources easily accessible on the UNM campus are the Clark Field Library, housed and staffed at the Department of Anthropology, and the main UNM libraries, which include the special collections in the former Anderson and Coronado rooms. OCA also has immediate access to the Maxwell Museum of Anthropology and Museum of Southwestern Biology collections for comparative study, and maintains close ties with teaching and research faculty and staff in the UNM Geology and Biology departments for consultation and project involvement.

Curation
OCA holds Curation Agreements with the Maxwell Museum of Anthropology, University of New Mexico; the Centennial Museum at The University of Texas at El Paso; the Museum of New Mexico in Santa Fe; and the Texas Archeological Research Laboratory at the University of Texas at Austin. OCA maintains a continually updated set of procedures for submission of collections for each of these institutions, and verifies their applicability at the outset of each project where curation is anticipated.

Past Performance
OCA has undertaken numerous projects for federal and state agencies, municipalities, and private firms. Many of these have been performed under multi-year, multiple task open-end contracts with agencies such as the New Mexico Bureau of Land Management, U.S. Army Corps of Engineers, Albuquerque District; U.S. Army Engineer District, Fort Worth; New Mexico State Highway and Transportation Department; and the Museum of New Mexico. Other large-scale projects include survey, excavation, analysis, and Native American consultation for Transwestern Pipeline Company and the Mid America Pipeline Company.

2

THE POLL, PART ONE

There can be no initiative without reason to fuel it. That's why the SCI FI Channel, as part of its public advocacy efforts and planned documentary of the Roswell Incident, commissioned RoperASW, a leading marketing intelligence firm, to poll Americans' beliefs regarding UFOs and extraterrestrial life. In late August 2002, RoperASW conducted the first of two phone surveys via its weekly OmniTel service, designed to obtain measures among a nationally representative sample of adults.

UFOs & Extraterrestrial Life

Americans' Beliefs and Personal Experiences

Prepared for the SCI FI Channel

September 2002

Roper Number: C205-008232

Table of Contents

I. Methodology

This study was conducted by RoperASW via OmniTel, a weekly national telephone omnibus service. The sample consists of 1,021 male and female adults (in approximately equal number), all 18 years of age and over.

The telephone interviews were conducted from August 23rd through August 25th, 2002, using a Random Digit Dialing (RDD) probability sample of all telephone households in the continental United States.

Interviews were weighted by five demographic factors: age, sex, education, race and geographic region. Weights were applied to ensure accurate representation of the adult population in each of these areas.

The margin of error for the total sample is +/- 3%.

II. Highlights

Most Americans Psychologically and Spiritually Prepared for Proof of Extraterrestrial Life

Most Americans appear comfortable with and even excited about the thought of the discovery of extraterrestrial life. Three-quarters of the public claim they are at least somewhat psychologically prepared for the discovery of extraterrestrial life, and nearly half are very prepared.

Such a discovery would not be difficult for most Americans to reconcile with their religious beliefs. Should the government make an announcement about the discovery of extraterrestrial life, only a very small proportion expect it to change their religious beliefs at all.

Additionally, slightly over half of Americans are at least somewhat interested in encountering extraterrestrial life forms themselves here on Earth. This is particularly true of males and of 18–24-year-olds.

Government Knows More Than It Is Telling

In the view of many adults (55%), the government does not share enough information with the public in general. An even greater proportion (roughly seven in ten) thinks that the government does not tell us everything it knows about extraterrestrial life and UFOs. The younger the age, the stronger the belief that the government is withholding information about these topics.

This is not a situation that most Americans would like to have continue. Provided national security is not at risk, most believe that the government *should* share information it has about other intelligent life and UFOs with the public. Males and adults below the age of 65 are more inclined to support the declassification of government information relating to such phenomena. Naturally, those with a belief and an interest in extraterrestrial life are also more likely proponents of revealing such information to the public.

The Alien Next Door?

Perhaps Americans expect to take a government announcement about extraterrestrial life in stride because many Americans already believe in the extraterrestrial. Two-thirds of Americans say they think there are other forms of intelligent life in the universe, and nearly half say they believe that UFOs have visited the earth in some form over the years (48%) or that aliens have monitored life on earth (45%). In fact, more than one in three Americans (37%) believe that humans have already interacted with extraterrestrial life forms. These beliefs tend to be more prevalent among males and among adults under the age of 65.

When it comes to alien abductions, one in five Americans in general and over half (57%) of those who say that humans have already interacted with extraterrestrial life believe that abductions have taken place. Once again, males and 18–64-year-olds are most likely to hold such a belief.

Alien Encounters

One in seven Americans say that they or someone they know has had at least one encounter of the "First," "Second," or "Third" kind. The largest proportion (12%) say they or someone else they know has seen a UFO at close quarters. Much smaller proportions say they or an acquaintance has seen a UFO cause a physical effect on objects, animals, or humans (3%), or has had an encounter with extraterrestrial life (2%). Among those who believe in abductions, one-third claim to have experienced, or know someone who experienced, a Close Encounter of their own.

When it comes to other unusual personal experiences, 1.4%, or 2.9 million Americans, say they have experienced at least four of five key events that believers of UFO abductions have identified as being of particular interest in examining whether UFO abductions may have actually taken place. Perhaps not surprisingly, those who believe in abductions and who have experienced, or know someone who experienced, a Close Encounter are more inclined to report an occurrence of at least four such events.

III. Most Americans Psychologically Prepared for Proof of Extraterrestrial Life

Three in Four (74%) Claim They Are at Least Somewhat Psychologically Prepared for an Official Government Announcement Regarding the Discovery of Intelligent Extraterrestrial Life

Clearly, a majority of Americans are ready for the discovery of extraterrestrial life, with 42% saying they are "very prepared" and 32% saying they are "somewhat prepared." Psychological preparedness is particularly common among the following demographic groups:

- Males (83%)
- 18–64-year-olds (79%)
- Those with household incomes of $50,000 or more (85%)
- Residents of the West (82%)
- Those with Internet access at home (82%)

% Very/Somewhat Psychologically Prepared

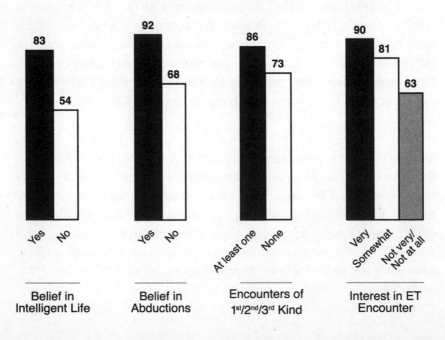

| Belief in Intelligent Life | Belief in Abductions | Encounters of 1st/2nd/3rd Kind | Interest in ET Encounter |

Americans who believe in extraterrestrial phenomena and who claim to have had, or know someone who had, a Close Encounter of their own are also significantly more likely to say they are psychologically prepared for the discovery of intelligent life.

Announcement of Extraterrestrial Life Would Not Prompt a Religious Crisis

Very few Americans say that an official government announcement about the discovery of intelligent extraterrestrial life would cause them to question their religious beliefs. A full 88% say that such an announcement would have no impact on their religious beliefs.

As age increases, so does the likelihood that religious beliefs will not be changed by the discovery of extraterrestrial life. Those 50 years of age and over are significantly more likely than 18–24-year-olds to claim that such a discovery would have no impact on their beliefs at all.

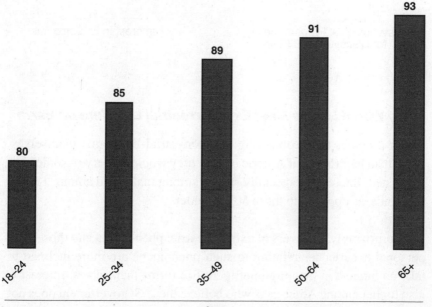

% Whose Religious Beliefs Would Not Change at All

Age	%
18-24	80
25-34	85
35-49	89
50-64	91
65+	93

Age

Those who say that they are "very psychologically prepared" for the discovery of extraterrestrial life are more inclined to believe that their religious beliefs will not be affected by such a discovery. However, the reverse is true of those who are "very" interested in encountering an extraterrestrial themselves—they are less likely to state that their religious beliefs would not be changed by the discovery of intelligent life.

% Whose Religious Beliefs Would Not Change at All

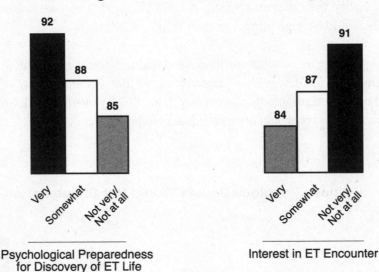

| Psychological Preparedness for Discovery of ET Life | Interest in ET Encounter |

Many Would Like to Meet Extraterrestrial Life Here on Earth

When it comes to encountering extraterrestrial life forms themselves, more than half (52%) of Americans say they would be very or somewhat interested. Interest is especially strong among males and among 18–24-year-olds, as opposed to those 50 and older.

Not surprisingly, believers of extraterrestrial phenomena and those with personal experiences relating to such phenomena are more inclined to take an interest in meeting intelligent life forms themselves. Interest is also higher among Americans who believe the U.S. government does not share enough information with the public and does not tell us everything it knows about extraterrestrial life and UFOs.

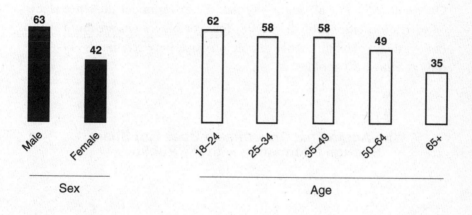

% Very/Somewhat Interested in Encountering Extraterrestrial Life

---

% Very/Somewhat Interested Among Those Who . . .

- Believe in intelligent life: 64%
- Believe in abductions: 83%
- Have had, or know someone who had, a Close Encounter: 79%
- Feel they are "very psychologically prepared" for the discovery of extraterrestrial life: 67%
- Think the government does not share enough information with the public: 58%
- Believe the government does not tell us everything it knows about extraterrestrial life and UFOs: 58%

IV. Many Wonder What Government Really Knows about Other Intelligent Life

The Government Has More Information Than It Is Sharing with the Public . . .

Over half (55%) of all adults say that the government does not share enough information with the public. Younger Americans are more apt to believe this; a clear difference of opinion exists between the 18–49-year-olds and those 50 or older.

% Who Agree That Government Does Not Share Enough Information with the Public

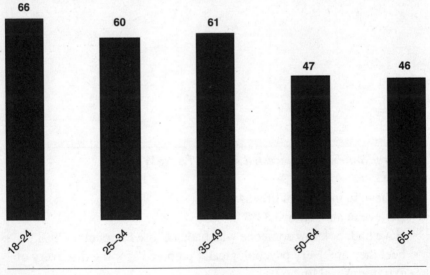

Age

Individuals with an interest and belief in—not to mention personal experience with—extraterrestrial or UFO encounters feel that the government has more information than it is sharing.

Most Americans (53%) say that their level of trust in the U.S. government has remained stable over the past five years. Three in ten (29%)
claim they trust the government less than they did five years ago.

**% Who Agree That Government Does Not Share
Enough Information with the Public**

59	63	50
Very	Somewhat	Not very/Not at all

Interest in ET Encounter

64	53
Yes	No

Belief in Abductions

66	54
At least one	None

Encounters of
1st/2nd/3rd Kind

... Particularly with Respect to Extraterrestrial Life and UFOs

Over two-thirds of Americans say that the government is not telling the public everything it knows about UFO activity (72%) or extraterrestrial life (68%).

The younger the adult, the more likely he/she is to believe that the government is withholding information about UFOs or other intelligent life.

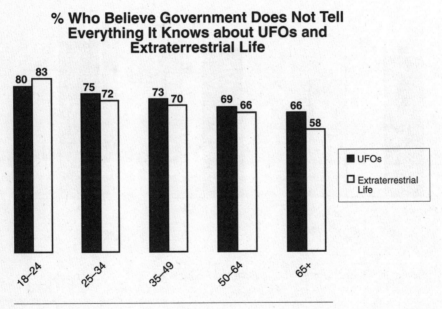

% Who Believe Government Does Not Tell Everything It Knows about UFOs and Extraterrestrial Life

Not surprisingly, those who believe and take an interest in extraterrestrial life and UFO activity are significantly more inclined to believe that the government does not share everything it knows about these topics.

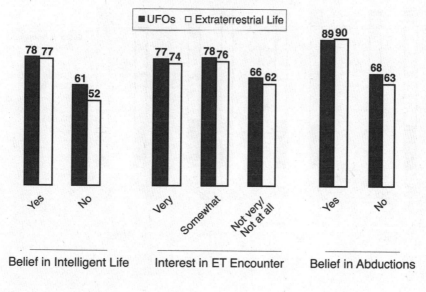

% Who Believe Government Does Not Tell Everything It Knows about UFOs and Extraterrestrial Life

■ UFOs ☐ Extraterrestrial Life

Belief in Intelligent Life — Interest in ET Encounter — Belief in Abductions

The Government Should Be Sharing This Information with the Public

Six in ten adults believe that information regarding UFO sightings (60%) and extraterrestrial life (58%) should be declassified if national security is not at risk. Another one in ten says it depends upon the situation.

Opinions vary depending on sex and age. Males are significantly more inclined to say that such information should not be kept classified, whereas females are more likely than males to claim they don't know or it depends on the situation.

With respect to age, adults 65 or older are significantly less inclined to believe that information regarding UFOs and extraterrestrial life should be shared with the public. Compared to its younger counterparts, this group is more likely to decline a set opinion on the topic.

Classification of Government Information Relating To . . .

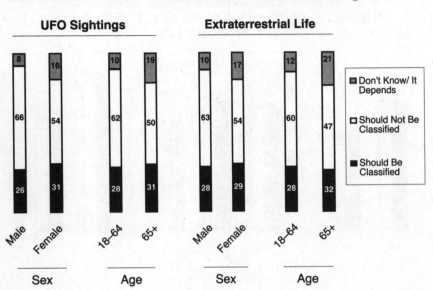

Adults who consider themselves at least somewhat psychologically prepared for the discovery of extraterrestrial life are significantly more inclined to believe that government information regarding this and UFO
sightings should not be kept secret from the public. This same opinion is
shared by believers of extraterrestrial life as well as those who are very
interested in encountering such life forms themselves.

% Who Believe in Declassifying Government Information Relating to UFO Sightings and Extraterrestrial Life

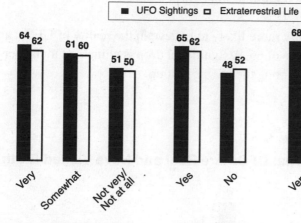

■ UFO Sightings □ Extraterrestrial Life

| Psychological Preparedness for Discovery of ET Life | Belief in Intelligent Life | Interest in ET Encounter |

V. A Majority Already Believe in the Extraterrestrial

To Many Americans, UFOs Are Real and Have Visited Earth in Some Form

Over half (56%) of the American public think that UFOs are something real and not just in people's imagination. Nearly as many (48%) believe that UFOs have visited Earth in some form.

Males are significantly more likely to believe in the reality of UFOs, as are those under the age of 65. A significant drop is witnessed in the percentage of believers among the 65+ age group.

% Who Believe That UFOs Are Real and Have Visited Earth

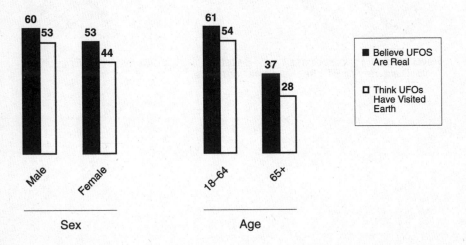

In terms of their psychological profile, a majority of adults who hold the beliefs and opinions outlined below also fall into the category of UFO believers.

Among adults who . . .	% Think UFOs Are Real	% Believe UFOs Have Visited Earth
Believe government does not share enough information with the public	60%	54%
Believe government does not tell us everything about extraterrestrial life and UFOs	70%	62%
Are very/somewhat psychologically prepared for the discovery of extraterrestrial life	63%	55%
Believe in intelligent life	73%	65%
Are very/somewhat interested in an ET encounter	72%	65%
Believe in abductions	93%	90%
Have experienced (or know someone who experienced) a Close Encounter	82%	76%

Humans Are Not the Only Form of Intelligent Life in the Universe

Two-thirds (67%) of adults think there are other forms of intelligent life in the universe. This belief tends to be more prevalent among males, adults ages 64 or younger, and residents of the Northeast as opposed to North Central and South.

% Who Think There Are Other Forms of Intelligent Life in the Universe

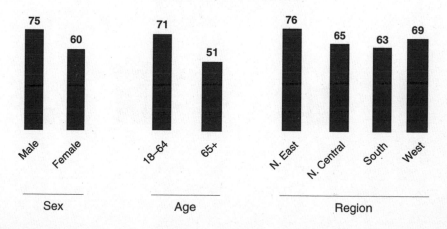

Moreover, nearly half (45%) of adults believe intelligent life from other worlds has monitored life on Earth. This belief is again held by a significantly higher percentage of males (50% vs. 40% of females) and of adults ages 64 or younger (49% vs. 28% of 65+ year olds). The belief is especially strong among 18–24-year-olds (59%).

The same patterns are evident when analyzing the belief that humans have already interacted with intelligent life from other worlds. While over one-third (37%) of all adults can be considered believers of this phenomenon, the percentage drops significantly among the 65+ age group (16% vs. 42% of 18–64-year-olds). Females are also less inclined to count themselves among the believers (33% vs. 40% of males).

Taken?

One in five Americans in general (21%) and over half (57%) of those who say that humans have already interacted with extraterrestrial life believe that humans have ever been taken or abducted by other intelligent life forms.

Once again, differences of opinion exist between the genders and age groups.

% Who Believe in Abductions

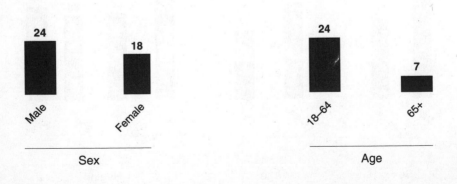

A belief in extraterrestrial life and abductions by such life forms is more prevalent among those who feel the government is hiding relevant information; who are very psychologically prepared for a discovery and themselves very interested in an extraterrestrial encounter; and who already have personal experience with a Close Encounter.

Among adults who . . .	% Think There Are Other Forms of Intelligent Life	% Believe in Abductions
Believe government does not tell us everything about extraterrestrial life and UFOs	76%	28%
Are very psychologically prepared for the discovery of extraterrestrial life	79%	32%
Are very interested in an extraterrestrial encounter	84%	40%
Have experienced (or know someone who experienced) a Close Encounter	83%	47%

VI. UFO Experiences Are Not Foreign to Americans

One in Seven Americans Say They or Someone They Know Has Had an Experience Involving a UFO

A total of 14% have had or know someone who has had at least one encounter of the "First," "Second," or "Third" kind. The largest proportion (12%) say they or someone else they know has seen a UFO at close quarters. Much smaller proportions say they or someone they know has seen a UFO cause a physical effect on objects, animals, or humans (3%) or has had an encounter with extraterrestrial life (2%).

Claims of at least one Close Encounter are more common among males and among 18–24-year-olds.

% Who Say They or Someone They Know Has Experienced at Least One Close Encounter

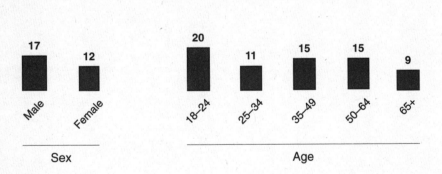

17	12	20	11	15	15	9

Male · Female — Sex

18-24 · 25-34 · 35-49 · 50-64 · 65+ — Age

Of those who believe in intelligent life, 17% claim to have had (or know someone who had) an encounter with a UFO or extraterrestrial life. This percentage nearly doubles when looking at believers of abductions—32% of these individuals state that they or someone they know has had at least one Close Encounter.

The higher the interest level and psychological preparedness of the individual, the greater the likelihood that he/she has—or knows someone who has—personally experienced such phenomena in the form of a Close Encounter.

% Who Say They or Someone They Know Has Experienced at Least One Close Encounter among Those Who . . .

- Are very interested in an extraterrestrial encounter: 28%
- Feel they are very or somewhat psychologically prepared for the discovery of extraterrestrial life: 16%
- Think the government does not share enough information with the public: 17%
- Believe the government does not tell us everything it knows about extraterrestrial life and UFOs: 16%

Four in Ten Americans Have Experienced at Least One of Five Events Associated with UFO Phenomena

While a number of unusual personal experiences arguably can be tied to UFO phenomena, believers of UFO abductions have identified five events that are of particular interest in examining whether UFO abductions may have actually taken place. These five events are highlighted in the list of unusual personal experiences shown below, and hereafter will be referred to as "key" events.

"Waking up paralyzed with a sense of a strange person or presence or something else in the room" tops the list of events that have ever been experienced. 18–64-year-olds are significantly more likely to make this claim than are those ages 65 and older (22% vs. 13%, respectively).

Event	% Ever Experienced
Waking up paralyzed with a sense of a strange person or presence or something else in the room	20
Seeing a ghost	15
Feeling that you were flying through the air although you didn't know why or how	14
Feeling that you left your body	13
Seeing a strange figure—a monster, a devil, or some other frightening being—in your bedroom or closet or somewhere else	13
Finding puzzling scars on your body and neither you nor anyone else remembering how you received them or where you got them	12
Seeing unusual lights or balls of light in a room without knowing what was causing them or where they came from	10
Experiencing a period of time of an hour or more, in which you were apparently lost, but neither you nor anyone else could remember why or where you had been	8
Having vivid dreams about UFOs	5

In our analysis of the five "key" events, adults who responded affirmatively to a control question—"Heard or saw the word TRI-NERVER and knew that it had a secret meaning for you"—were excluded from the base of respondents. Only 1.2% of all adults claimed that this statement applies to them.

Four in ten (40%) Americans say that they have experienced at least one of the five key events. This is especially true of males, 18–24-year-olds, and individuals with incomes below $50,000.

The likelihood of having experienced at least one key event is significantly higher among those who believe in intelligent life and abductions, take an interest in extraterrestrial encounters, and think the government knows more than it's sharing.

% Who Have Experienced at Least One Key Event

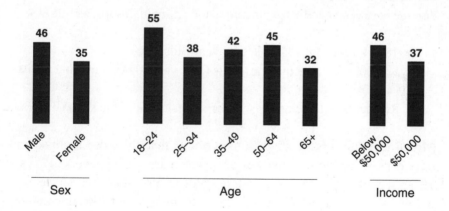

> *% Who Say They Have Experienced at Least One Key Event*
> *among Those Who . . .*
>
> • Believe in intelligent life: 46%
> • Are very/somewhat interested in an extraterrestrial encounter: 50%
> • Believe in abductions: 54%
> • Have had, or know someone who had, a Close Encounter: 58%
> • Think the government does not share enough information with the public: 47%
> • Believe the government does not tell us everything it knows about extraterrestrial life and UFOs: 47%

1.4%, or 2.9 Million Americans, Have Experienced at Least Four of Five Key Events

Certain groups of individuals are more likely to claim that they have experienced at least four of the five key events. Significant differences emerge between specific age, income, and regional groups, as follows:

• 18–24-year-olds (3%) vs. 65+-year-olds (0%)
• $20,000–$29,999 income adults (3%) vs. $50,000+ income adults (<1%)
• Residents of the Northeast (2%) vs. residents of the West (0%)

Moreover, believers of abductions are significantly more inclined to have experienced four of five key events than are non-believers (4% vs. 1%, respectively). This same finding holds true for those who have experienced, or know someone who has experienced, a Close Encounter (5%) versus those who have not (1%).

DECLASSIFIED

3

THE DIG DIARIES

Months of planning finally came to fruition in mid-September, 2002—SCI FI, in partnership with the University of New Mexico's Office of Contract Archeology, set out to uncover conclusive physical evidence that could establish the 1947 "Roswell Incident" as science fiction or science fact. For nine days in September, an excavation at the J. B. Foster Ranch would require many hands under the project leadership of OCA principal investigator William Doleman, including assistant archeologists J. Robert Estes and Louis Romero, and volunteers Alana Lynn Andrews, Todd Fischer, Harold Granek, John and Nancy Johnston, Kate Leinster, Jerry Lowe, Glenn Porzig, Debbie Ziegelmeyer, and Chuck Zukowski. Then there were the people who got their hands dirty on other aspects of the project, including technical advisers Thomas Carey and Donald Schmitt; SCI FI's director of special projects, Larry Landsman, and senior vice president, programming, Thomas P. Vitale; Sunbelt Geophysics' David Hyndman and Sid Brandwein; backhoe operator Elgio "Alley Cat" Aragon; independent freelance journalist Leslie Kean; freelance photographer Babak Dowlatshahi; and the MPH Entertainment crew that filmed the dig for SCI FI's November 22 documentary, *The Roswell Crash: Startling New Evidence.*

In keeping with a nondisclosure agreement that prevented them from discussing any details prior to airing the special, several participants kept personal daily journals, segments of which appeared on SCIFI.COM. Here, for the first time, are the complete "Dig Diaries" of William H. Doleman, Thomas Carey, and Donald R. Schmitt (who also preface several days

Members of the September 2002 Roswell dig pose for an official commemorative photograph. Kneeling is University of New Mexico's principal investigator, William Doleman. From left to right, standing: UNM's Robert Estes, Jerry Lowe, UNM's Louis Romero, Alana Lynn Andrews, Kate Leinster, John Johnston, Harold Granek, Debbie Ziegelmeyer, Chuck Zukowski, and Nancy Easley Johnston.

of these entries), Thomas P. Vitale, and volunteer dig-team members Kate Leinster, Debbie Ziegelmeyer, and Chuck Zukowski.

September 13

Thomas J. Carey, Roswell investigator, coauthor of
The Roswell Report

As usual, I let packing my bags go to the last minute because I hate packing my bags; I keep trying to pack something to cover every possible contingency, and for this trip that's saying something.

An archeological dig in the high New Mexico desert, looking for pieces of a crashed flying saucer. Who would have thought we could pull it off? But here we go—after 13 years of talk and disappointment, it's actually happening. I finish packing my bags at 1:30 A.M. Time to get some sleep.

September 14

Thomas J. Carey

Up at 4:30 A.M. No breakfast. Off to the airport to catch my 7 A.M. flight. It's still pitch-black outside, but no traffic on I-95, which is good. I get a chance to drive Doreen's brand-new Toyota SUV to the airport. Smooth ride all the way. At Philadelphia International Airport, there is a maze of new detour signs directing traffic around new construction. Every time I go to the airport, there seems to be "new construction." Doreen leaves me off in front of the America West Terminal, and we say our goodbyes. This is nothing new to us, as this will be my fifteenth trip to New Mexico since 1998.

I am flying to Albuquerque by way of Phoenix. I will meet Don in Albuquerque, where we will pick up a rental car at the airport. Don's plane gets in about the same time as mine, so there should be no missed connections. We plan to spend the evening at Nate Twining's (Nathan Twining Jr., son of the former Air Force general and chairman of the Joint Chiefs of Staff in the 1950s).

Thank God Don's friend Chester Lytle won't be joining us for dinner. He talks so slowly that, when I am so tired from the flight, it becomes painful trying to stay awake. Last time at Nate's, I was literally out on my feet. Don, on the other hand, needs and thrives on only four hours sleep. Though he has been known to fall asleep at the most inopportune times—in the middle of someone talking to him, or while operating a slide projector for the speaker in front of a large audience.

The flights themselves are uneventful and pretty much on time. The flight from Philly to Phoenix is only half-full, so I stretch out and try to catch up on some sleep; I will need it for dinner at Nate's, which can go on for hours. The short flight from Phoenix to Albuquerque is jammed—not a seat to be had—which seems to be the norm for airline travel these days, with their misguided "hub concept." Reminds me of the old TV show *Zoo Parade.*

8 P.M.

Dinner with Nate and other guests at his mansion in Albuquerque. Always a great menu, as Nate has two live-in cooks on the premises. Tonight there are two main dishes: fish and steak. I take the steak, of course. Don and I tell some Roswell stories while the other guests hang on our every word. Nate tells some UFO stories of his own—and he has a lot—along with some reminiscences of his famous father.

His father, of course, knew all about Roswell and the UFO, but never told Nate or the family what he knew. For his part, Nate is an internationally known pianist, but makes his living from real estate, oil and art. After dinner, we adjourn to the family room with the gigantic TV. Nate, a boxing fan, has taped the Oscar De La Hoya vs. Fernando Vargas championship fight and plays it on the tube.

A health aficionado as well, Nate wants to show us a new electronic device he just bought that is supposed to cure all ills by varying the sequences of the electronic pulses it emits. I volunteer to test it out, which is a mistake. Nate sets it to "Needs a Lot of Help" and turns it on. I tell him the electronic pulses are very faint, so he increases their intensity. Just then, I start to come out of my chair; this is what being electrocuted must feel like. I scream out to turn it down, while Schmitt laughs away in the other chair. After the intensity is lowered to a comfortable level, I stay on the machine for about a half-hour. Feeling lucky to be alive, I tell Nate that I feel better as I unhook myself and exhort Schmitt to give it a try. He doesn't.

2 A.M.

I can't hold out any longer, so I excuse myself and turn in. Schmitt and Nate are still going strong.

September 15

Thomas J. Carey

Up at 7 A.M., as I have not yet adjusted to the two-hour time difference between Pennsylvania and New Mexico. It usually shows up the first two or three evenings that I am in New Mexico. It doesn't seem to bother Schmitt, though, as there is only a one-hour time difference between Wisconsin and New Mexico . . . not to mention the fact that he needs only four hours of sleep, anyway.

We had hoped to see retired Lt. Col. George Johnston this morning or afternoon in Albuquerque and persuade him to be interviewed for the SCI FI documentary on Roswell. Col. Johnston is a firsthand eyewitness to a piece of "memory metal." Back in 1952, he had been shown a piece of it by his commanding officer while stationed at Holloman AFB in Alamogordo, New Mexico. When I interviewed Col. Johnston several months ago, he had said to come see him whenever I was in Albuquerque. When I called him just prior to this trip, however, he said that yesterday was no good because he would be watching the football games on TV, and that today was also no good because he would be in church all morning and eating out in the afternoon. We run into this all the time—unkept promises and false commitments. I mean, it's not like we're in town all the time, but people will not alter their schedules to meet with us.

We had also wanted to meet with a fellow from Albuquerque named Jay Miller, who had certain information regarding the Corona crash site. We had been dealing through a third party in Socorro, New Mexico, to try getting Jay to talk with us. I had sent an e-mail to our Socorro contact to set something up for yesterday or today; any time would be OK, I said. To no avail, however, as we hear nothing further from Jay or our contact. This, again, is more or less par for the course, as it is extremely difficult to get people to do anything for us, even when they have promised to help. Most just want to know the latest information regarding Roswell and use the promise of assistance as the "hook" to obtain it. They also feel that the moment they fulfill their promise we won't need them anymore, but as long as it remains unfulfilled, we will still need and appreciate them.

Still full from last evening's repast, we don't stop for breakfast and head out around 8:30 A.M. for Roswell. Since this is Don's vehicle, he is the driver. Don has only one good eye, as the other was severely injured in an auto accident years ago. My heart is in my throat again, because Don tends to see things—such as curves in the road—at the last moment. As a result, my life has passed before me on many occasions in New Mexico, especially driving on the curvy, unlit two-lane road from Corona to Roswell late at night.

We stop at Allsup's in Vaughn for a snack to hold us until dinnertime. Vaughn is the halfway point in the three-hour drive from Albuquerque to Roswell.

NOON

We arrive in Roswell and head straight for the UFO Museum on Main Street. The museum is our base of operations whenever we're in

Roswell. Don already has some phone messages waiting for him upon our arrival. I never get any. Besides our research on the case, Don is also on the museum's board of directors and, as a result, has additional business to transact. There always seems to be something going on there that requires his attention. Because of this, I have scheduled my own car-rental while in Roswell, so that I don't have to depend on Don for transportation. This hopefully will ameliorate problems that have occasionally popped up during past trips.

6 P.M.

Don's business at the museum, as usual, has taken more time than originally planned. It is dinnertime before he is finished. We adjourn to the Golden Corral for dinner, where we have eaten many, many times on previous trips. I like the mashed potatoes and gravy there; Don likes the salad bar.

7:30 P.M.

Back to the museum after dinner. Don has to finish up our next "Roswell Report" column and submit it to the SCI FI Channel today for posting on their website.

9 P.M.

I excuse myself to go to the bathroom at the far end of the museum. It is now dark outside—and inside as well; all the lights are out except in the computer room. Halfway into my "stroll" to the bathroom, things suddenly become eerie. I can't explain it, but I am spooked walking by myself in this big, dark museum dedicated to aliens. After finishing in the bathroom, I literally sprint back to the computer room, still not sure what is going on. Noticing my pale appearance and rapid breathing, Don's curiosity is piqued. I tell him about what just happened, and he proceeds to tell me that the museum is said to be "haunted." Apparently, several members of the museum staff have reported strange noises throughout the museum—specifically, the sound of children playing upstairs after hours. The museum director, Julie Shuster, reported seeing a white floating "apparition" one evening, right in the area of the museum to which I had gone. Great. I tell Don in no uncertain terms, "Any strange noises, and I'm outta here!"

10 P.M.

After Don finishes the article, we finally check into the Holiday Inn Express, where we will be staying for the duration of our trip. This is a big step up for us since, during our last trip to Roswell, we stayed at the Roswell Budget Inn La Cucaracha, down the street and across the tracks. There, the cockroaches wear nametags. (Don had justified the choice of lodgings by saying, "The price was right.")

September 16

Dr. Bill Doleman, Principal Investigator and Project Director, University of New Mexico Office of Contract Archeology

I leave Albuquerque at 11:30 A.M. after picking up Todd Fischer at the airport. Meanwhile, my crew—Lou Romero and Bob Estes—head straight to Corona to meet with Don Schmitt and Tom Carey, his research associate. We meet up with them and pick up some snacks at the gas station before proceeding out to the Foster Ranch site. We drive past the Hines House, into the project area, and past the location I had been shown before. Don gets his bearings and we drive to the actual site. I think I later succeed in confirming it as the same location depicted as the "debris field" in Schmitt and Kevin Randle's 1994 book, *The Truth about the UFO Crash at Roswell*.

Don takes us to the stone cairn and then to the initial impact point, which he had been shown by more than one eyewitness. With Don's help, we mark the centerline of our grid system along the centerline of the no-longer-visible furrow. My assistant Bob walks out about 200 meters, then back and forth until Don says something like, "There. He's on line there."

Knowing that we had to rely entirely on Don and Tom's knowledge of the site, I ask how much uncertainty there is in the location and orientation of the furrow. They seem sure about the initial impact point, orientation and length, but add there might be as much as 200 feet—about 60 meters—of uncertainty regarding the centerline's location. Using the centerline they showed us, we start setting up a 300-meter by 120-meter grid system; Don, Tom and Todd head into Roswell to catch some sleep before coming back to guard the site overnight.

Before heading to Corona for the night, we unload all the equipment we think will be safe on the ground in the middle of nowhere. Corona is actually the nearest town with a motel—a 45-minute drive versus two hours to Roswell. In many respects, it should be called the *Corona* crash site!

Thomas J. Carey

9 A.M.

Off to pick up my rental car at Avis, located at the airport, which itself is located on the former Roswell AAF base south of Roswell. From past experience, it's best that Don and I have separate vehicles to prevent my being inconvenienced by Don's ever-changing schedule or the endless debates about who is going where and when and which is more important. I was once unexpectedly stranded by Don at the airport in Albuquerque, with no vehicle and no place to go, when he realized that his publicist had scheduled him to fly home a day before my flight out. This just strikes me as being an unnecessary waste of time for everyone, as well as an irritation for me.

9:15 A.M.

It's off to the Roswell UFO Museum to set up shop for the remainder of our stay. The museum director, Julie Shuster (who is also the daughter of Walter Haut, the 1st Lieutenant who wrote and distributed the 1947 press release that Roswell Army Air Field had captured a downed flying saucer), is recently back from major surgery and due to start chemotherapy in a week or so. She has been very helpful to our investigation, and to the case in general, by dealing with all of the B.S. that comes into the museum and across her desk. There are also dark forces lurking in the shadows who would like nothing better than to remove Julie from her position in order to place a "friendly" at the museum to facilitate their agenda. So far, we have been able to withstand every assault from these conspirators.

Walter Haut, one of the museum founders, is in his office most afternoons during the week. He is 80 years old (give or take a couple of years) and gets around with difficulty due to bad knees. His wife, "Pete," passed away a few years ago.

Glenn Dennis, an embalmer back in 1947 who claims to have had a nurse friend involved in a preliminary autopsy of dead aliens at the

Roswell AAF base hospital, is also one of the museum founders. I am told he comes by the museum much less frequently than Haut, perhaps only two or three times a week. I have brought with me a few pictures of a black male, sent to me by a source who says that this man was his science teacher back in the 1960s. The teacher supposedly told him one day that he had been part of an autopsy team at Roswell in 1947 that did its work on a dead alien that had crashed near there.

Max Littel, the third founder of the museum, recently passed away. A real-estate developer by trade, he had wanted to move the museum outside of Roswell proper, but the plan was fortunately voted down amid much angst and acrimony.

I spend most of the day going over my notes and the names of alleged witnesses living in Roswell.

6 P.M.
Dinner at the Golden Corral again. Mashed potatoes and gravy. Ahhhh.

7 P.M.
Back to my hotel room. I have excused myself from any further activity because the Eagles are on *Monday Night Football* tonight against the Washington Redskins. I notice right away that Roswell has two things I don't have back home: two excellent "oldies" radio stations and the Turner Classic Movies cable station. I am already hooked on them. I watch the game and quietly fall asleep.

Donald R. Schmitt, Roswell investigator, coauthor of
The Roswell Report
I've had the opportunity to visit the Roswell debris-field site several times since February 1989. It was on one of those early trips that Mack Brazel's son Bill drove us out in his red and white Ford pickup and gave us a firsthand tour. Through the years, other witnesses have independently taken us to that very site, each adding their own personal recollections. But in all that time, I had never been out to this desolate region after dark. That's why I was especially excited to spend four nights doing security there.

Our first impressions of the site provide us with the feeling of total isolation, and of being placed on terrain that's more befitting of man's first steps on another planet. The fall of night truly amplifies this feeling. The utter lack of human presence is palpable, adding to the overall effect of the rest of the world being figuratively a thousand miles away. One cannot help but remember what had occurred after dark, that fateful night back in July 1947. The local rancher described a wicked storm that night, yet my experience at the site is that the stars have never looked so bright.

September 17

From the crash site "where it all began," the team began its full-scale archeological dig. Using standard archeological field methods of excavation, the hope was to find physical evidence of whatever crashed to Earth at the site back in 1947. By a process of elimination, the alternate possibilities have come down to just two competing choices: a rubber, balsa wood, and tinfoil balloon array, or a UFO.

Before the volunteers arrived for the actual digging, the site needed to be prepared. This involved mapping the site, then setting up a grid system to control the digging and properly record what was found. Areas of high potential, such as the alleged "500-foot gouge" which formed the "spine" of the grid system, were identified and marked with flags. The site was also secured to prevent it from being disturbed in any manner for the entire time of the dig.

Dr. Bill Doleman

My crew and I spend the entire day setting up the grid system. This entails using a transit, a stadia rod and 100-meter tape to place 10-inch nails in the ground every 10 meters along the baseline (coincident with the furrow's reported centerline). Ten-meter intervals are used because the geophysical prospection would use the same interval.

We take elevation readings that are eventually entered into a computer database, converted to rectangular coordinates and used in a GIS [Geographic Information System] to create a three-dimensional contour map of the site. The grid system we've laid out provides locational control for both our test excavations as well as for the geophysical prospection surveys. Thus, we're able to overlay our excavations, the geophysical prospection results and the contour map.

The University of New Mexico establishes a comprehensive grid system on September 17, 2002, based on the location of the "furrow" as provided by investigators Don Schmitt and Tom Carey. Under the direction of UNM's principal investigator, William Doleman, and his assistant archeologists, Bob Estes (pictured) and Louis Romero, elevation readings were recorded and used in a GIS system to create a three-dimensional contour map of the site. The final grid system measures approximately 300 by 120 meters.

Thomas J. Carey

9 A.M.

Don and I are off to the debris field to meet the archeologists from the University of New Mexico. They will be our professional, expert advisers for the planned survey and excavation of the location where everyone agrees—investigators and U.S. Air Force alike—that something came to ground in 1947. The argument is what that "something" was. We have eliminated everything except a UFO. The Air Force has eliminated everything except rubber weather balloons, tinfoil and radar targets from the defunct "Project Mogul" (a 1947 attempt to have constant-level, high-altitude, balloon-borne acoustic sensors detect sound waves from the then-expected detonation of an atomic bomb by the Soviet Union).

NOON

We meet the archeologists in Corona—the same town where Mack Brazel visited just before traveling to Roswell to report his find. The lead archeologist, Dr. Bill Doleman, has, unfortunately from my perspective, brought along the frenetic name-dropper himself, Todd Fischer—as ballast, I guess. Actually, Todd is one of Don's "hangers-on" whom he invites along on trips without my knowledge or consent. They don't really do anything but hope something good will happen to them by osmosis if they hang around Don and the Roswell case long enough. Don has promised me that Todd is here as one of the volunteer diggers. I will believe that when I see it.

At the debris site, Don points out the location of the alleged "gouge," and its direction and length. Doleman has members of his crew plant flagged stakes every 10 yards along its length. This is done to identify the area to be traversed by the ground-penetrating radar— actually called "Electronic Conductive" to try to obtain a "fingerprint" or evidence of the gouge from 1947. You can't see it now, as it has all been filled in by surface erosion and soil deposition, but I'm told that soil density studies should be able to detect it. We spend the rest of the afternoon at the site.

On the ride back to Roswell, Don tells me of his plan to "secure the site" overnight by "riding shotgun" in his SUV. To me, this seems to be unnecessary overkill. Let's face it—who knows where the place is, and who knows what's going on? But it's clear that Don's mind is made up and he is going to do it, come Hell or high water. The really bad news comes when Don volunteers my vehicle to Todd; "Junior G-Man" has offered to be Don's "graveyard-shift" relief, but has no way of getting out to the site.

7 P.M.

Dinner at the Golden Corral for the third straight night.

9 P.M.

Back to the hotel. I turn my vehicle's keys over to Todd. I will not get to use it for the remainder of this trip. So much for my schedule!

September 18

Site preparation continued with setting up the grid system, and an Albuquerque company had been brought in to conduct studies similar to those accomplished by ground-penetrating radar (GPR) in the hope of finding a "fingerprint" of the reported (but now hidden) gouge area. Such a finding would rule out any type of balloon as causing such a feature. The electromagnetic conductivity (EMC) survey procedure measures the differences in the conductivity of electrical current in soils of different densities due to differing moisture contents. A gouge that had been filled in by natural erosion over fifty-five years would presumably show greater electrical conductivity, containing more moisture than the surrounding soil because its makeup was less dense. The second phase of such "geophysical prospect-

Under UNM supervision, Dave Hyndman and Sid Brandwein of Sunbelt Geophysics begin geophysical prospection on September 18, 2002. Electromagnetic conductivity (EMC) and metal detection surveys are performed. The site's topography both reflects and controls many of the natural processes that have acted on its soils since 1947. Astonishingly, the tests reveal three anomalous conductivity "highs," two of which line up right next to the presumed crash furrow!

ing" was to conduct a search for any subsurface anomalies, such as metal fragments, within the grid area.

At this point, the unseasonably wet weather continued to threaten the entire project; in addition to making the two-track roads to the site almost impassable—a new way to get the large RV to the site had to be found—rain forced a premature termination of the day's planned activities at the site.

Dr. Bill Doleman

With the centerline baseline established, and each of the 301 nails tagged with a permanent tag containing the nails' grid system coordinates, we set up the transit at 100-meter intervals, turn right angles and tape out to nails marking the 60-meter boundary on each side of the baseline. We then set pinflags every 10 meters along the grid system boundaries, thus completing the 300- by 120-meter system.

About lunchtime, and right on schedule, Dave Hyndman and Sid Brandwein of Sunbelt Geophysics show up to begin geophysical prospection. They set right to it, and are almost done when a cold front we'd been watching all afternoon hits us with high winds and cold, driving rain. We hightail it out around 4:30 P.M., since we're not sure how the two-track roads will be when wet. Dave comes out about 30 minutes later. He and Sid head back to Albuquerque, while we go to the Willard Cantina for an excellent dinner.

Thomas J. Carey
8:30 A.M.
Complimentary breakfast at our hotel's breakfast bar. Todd, who mercifully is staying at another hotel, is somehow joining us for breakfast. He will do this for the remainder of the trip, much to my chagrin. We are thus subjected to his incessant babbling concerning every inane thought that enters his head. I feel like a sophomore in college again.

1 P.M.
Don, Todd, Larry Landsman of the SCI FI Channel and I head out to the debris field, to show Larry the lay of the land and see how the archeologists are doing. The GPR/EMC guy (whom I know only as Hyndman) started his scan of the debris field today in order to find

anything anomalous under the surface of the soil, such as the alleged "gouge" seen by several witnesses right after the crash of 1947. We introduce Larry to the archeologists and give him a tour of the site.

4 P.M.
Back to Roswell for dinner. Larry's buying, so we don't have to eat at the Golden Corral. We eat instead at Juana's, a fine Mexican restaurant located near the Roswell Super 8 motel, Kevin Randle's favorite (along with fast-food joint Church's Chicken).

11 P.M.
Back to the Holiday Inn Express, where I fall asleep watching a Turner Classic movie. Don has headed back to the debris field for his midnight–3 A.M. "shotgun" shift. Todd will relieve him at 3 A.M. . . . in my vehicle.

Donald R. Schmitt

My biggest concern is getting the production staff's RV to the debris-field site, since previous heavy rains have all but washed out many of the higher dirt and gravel trails. We've already had difficulties with our four-wheel drives; there's a lot more distance from axle to axle with an RV, and we hadn't yet found a safe route to prevent such a vehicle from either tipping or bottoming out.

We spend the better part of the afternoon discovering and creating a new dirt pack trail to the site. We mark it with flags, but know the true test for the actual camper won't come until Saturday.

September 19

Continuing marginal weather caused concern for the archeologists and geophysical prospectors. Although driving on the roads turned out better than expected, the wind made contour mapping of the entire site (as with any archeological excavation, all results—geophysical as well as archeological—were interpreted with the aid of the map) difficult, and affected the efficiency of the EC study, which is terminated slightly short of its goal. Meanwhile, back in Roswell, a key eyewitness to the 1947 events was interviewed for the first time on camera by MPH Entertainment.

Dr. Bill Doleman

Thursday dawns cool and drizzling, but shows signs of clearing. Dave and Sid are convinced that the project area roads won't be passable. We head out to the site anyway, given our tight schedule and the need to be ready for the volunteers on Saturday. The roads turn out to be perfectly fine.

We set to work, taking contour shots every five to 20 meters across the entire site. Our GIS wizard, Ron, will use them and some nifty software to interpolate contours. The more varied the site's terrain, the more closely we space the shots to ensure accuracy in the resulting topographic map. This map will be extremely important in interpreting the results of the geophysical prospection and test excavations, as the site's topography both reflects and controls many of the natural processes that have acted on its soils since 1947.

Although the day is more or less clear, the wind blows strong enough to be annoying. Contract archeologists in particular are used to working in unpleasant conditions, because their "field season" is all year long. Every New Mexico archeologist has an internal meter that measures discomfort, as well as lots of little tricks to minimize that discomfort, whether it be heat, cold, rain or snow.

The wind is the biggest enemy. This is because archeologists are always doing paperwork, which does not mix well with wind! My internal meter tells me it's just on the borderline between "annoying" and "obnoxious." These are levels we have to live with; when it gets to "intolerable" and then "impossible," it depends on how much of a hurry we're in. We continue mapping and complete 95 shots, about a third of the site.

Thomas J. Carey
8:30 A.M.

Complimentary breakfast at our motel. Unfortunately, Todd has somehow made it to breakfast with us. He should be sound asleep after playing "Junior G-Man" until the early morning hours. Go figure.

9:30 A.M.

Over to Pete Annaya's house to observe and help out with MPH Entertainment's videotaped interview of this important Roswell witness, who maintains that the late U.S. senator Joseph Montoya had a firsthand encounter with alien bodies in Hangar 84 on the Roswell base back in 1947. The witness's wife, Mary, corroborates his story, as does his son, who later worked for Montoya.

1 P.M.
After lunch, we head out to the debris field again to check on the archeologists and the electronic scans for the "gouge fingerprint" allegedly left by whatever crashed there in 1947. We also stake out a route to the site for the RV, to get it over rough terrain without bottoming out.

7 P.M.
Dinner at Juana's with the entire gang, including the production company (MPH Entertainment, who have arrived today), courtesy of Larry Landsman and his boss, Tom Vitale, director of programming at the SCI FI Channel.

Debbie Ziegelmeyer, independent UFO investigator
     I leave St. Louis at 8:30 P.M. to meet up with my brother, Chuck, in Colorado Springs. Unfortunately, my flight is delayed when the Colorado Springs airport is evacuated and closed due to a bomb threat. I finally arrive around 10 P.M.

September 20, 2002

A big day! With the volunteers arriving, all site preparations needed to be completed. The day started out with good news from geophysicist Dave Hyndman, whose electrical conductivity tests found anomalies in the area of the alleged gouge. The archeologists continued contour mapping of the site with alacrity, but were soon hampered by wind conditions. Using the EMC test results as a guide in conjunction with the grid system, sites where test pits would be sunk by volunteers the next day were located and marked (a process that became a bone of contention between Bill Doleman and some of the volunteers several days later).

Meanwhile, "location interviews" of key eyewitnesses to the 1947 events were conducted at the Roswell Industrial Air Center, formerly the top-secret airbase known as Roswell Army Air Field. It was here that the wreckage of whatever crashed in Mack Brazel's sheep pasture was brought, and where alien bodies were allegedly transported from the crash site. The location interviews took longer than expected, and were hampered by the sounds of aircraft taking off and landing.

Everyone—the archeologists, the MPH Entertainment and SCI FI Channel executives, and the volunteers (who hailed from Colorado, Texas, Arizona, England, and yes, Roswell)—converged at the UFO Museum and Research Center on Roswell's Main Street (the "base of operations" for the project) shortly after 5 P.M. for the big introduction. Unfortunately—and unexpectedly—about only half of the volunteers who had committed to the project actually showed up. This ultimately affected the amount of work that could be done during the time allotted for the dig, and forced some drastic action to be taken.

Dr. Bill Doleman

The day dawns clear and cool. It's going to be a long one.

After spending yesterday processing Wednesday's geophysical prospection data, Dave Hyndman drives down from Albuquerque to meet us at the Corona Motel and give us the results. On the phone, he said he had something "interesting" to show us. They had gotten about 90 percent of the site covered with an electromagnetic conductivity (EMC) survey, the first of two geophysical prospection techniques to be used on the site. The device that performs this survey, an EM-31, measures the conductivity in the ground by sending a radio signal into the ground and then measuring the strength and delay in the induced return signal. If the furrow is present but buried, it should show up as an anomaly, because the softer furrow fill will hold more moisture and thus exhibit greater conductivity than the more compact surrounding soils.

Dave's map showed three EMC "highs," one or two of which appeared to possibly correlate with local topographic features. Two of the highs were lined up right next to the presumed furrow axis, and of these one was definitely anomalous. This was pretty exciting stuff!

By the time we're done with Dave, it's almost 11 A.M. We hurry out to the site to continue our contour mapping. The wind picks up quickly; by 2 P.M. it's well into "obnoxious." Not only does it mess with the paperwork, but it makes the transit shake and my eyes water continuously.

I'm due in Roswell around 5 P.M. to brief the volunteer excavators, but the mapping we need to get done is going too slowly. We also have to start shooting in test pit locations (squares 50 centimeters on each side) so that they'll be ready for excavation by the volunteers tomorrow.

We work furiously despite the wind, but can't finish the contour mapping by the time we have to start shooting in the test pits. We use Dave's EMC survey map, which is tied to our grid system, to locate the EMC "highs," and shoot in three grids in each of the two "highs"

On September 20, 2002, University of New Mexico's principal investigator, William Doleman, briefs his staff of volunteer archeologists at the IUFOMRC (International UFO Museum and Research Center in Roswell). He explains that although much of the crash debris was apparently removed by the military within a few days of the impact, something may have been subsequently buried by natural processes, including erosion, deposition, and bioturbation. Doleman labels these potential artifacts as HMUOs, or "Historic Materials of Uncertain Origin." The lecture takes about an hour.

aligned with the furrow. We also shoot in one in the middle of the site, to provide us with baseline stratigraphic data—information on the nature of the site's "natural" soils, for comparison with whatever we find in the test pits.

At 4 P.M. I hightail it for Roswell, while Bob and Lou head back to Corona. I make it to the International UFO Museum and Research Center (IUFOMRC) by 5:15, thanks to a V8 and southeast New Mexico's long, straight roads.

Until now, the Foster Ranch project has been pretty much like any other fieldwork. All of that changes at the IUFOMRC, where there are SCI FI Channel folk and a contingent of people from MPH Entertainment of Burbank, California. Someone grabs me, hands me a cell phone, and tells me to inform somebody how to get to the site. Someone else needs me to explain things to a chopper pilot. A guy in a nice Hawaiian shirt (Miles from MPH) says I need to be "miked." Other folks are also asking questions.

It turns out that not all of the volunteers have arrived, and I can't start my briefing; good thing, as I've had no time to prepare my talk! Before we left Albuquerque, we had put together an information packet that described the basics of digging controlled holes in the ground, but it didn't provide an overview of the project. The packets, together with my lecture, would ease the process of making inexperienced volunteers into "instant field archeologists." But of course, I've forgotten to bring the packets with me! The museum staff has some Sharpies and a couple of poster boards ready, so I set about drawing pictures with lines, arrows and stick figures. By 7 P.M., everybody has arrived, the cameras are in place and the poster boards are covered with my scribbles.

Don, Tom and Larry introduce the project, and I go to work. I explain the overall goals of finding physical evidence of the crash in the form of leftover and now-buried debris, or debris in a buried furrow. Next I talk about how debris might have been collected or buried by animals, or might have fallen into now-sealed burrows, and how natural erosion processes might have obscured the furrow. Finally, after discussing the role of geophysical prospection (EMC and high-sensitivity metal detection) in finding both, I go over the digging methods we use in archeological testing, including the all-important concept of "provenience" (the "where things came from" question) and the specific methods used in the project to ensure site integrity. The lecture takes about an hour; by the time we're done, I'm exhausted. We all head out to find food and drink.

Thomas J. Carey
9:30 A.M.

I meet Marion Cox, 74, at the UFO Museum. Mr. Cox was an eyewitness to Air Force documents concerning the Roswell incident when he was stationed at Biggs AFB in El Paso, Texas, back in the early 1960s. Cox and I drive over to the Southeastern N.M. Historical Museum on Second Avenue, where for the rest of the morning we are interviewed by MPH Entertainment for our Roswell documentary. After the interview—which goes very well—everyone heads back to the Crash Down Diner (across the street from the UFO Museum) for lunch.

12:30 P.M.

Lunch at the Crash Down Diner with Marion Cox. Mr. Cox likes to talk about himself, but that's OK; after all, he did just give us a good interview on short notice.

AFTERNOON

Out to the airport, which used to be the Roswell Army Air Field back in 1947. MPH Entertainment will conduct on-site interviews of key 1947 witnesses, at the exact locations where they occurred. This was Don's and my idea; we insisted that this be done rather than just have more "talking head" interviews. Another Schmitt surprise rears its ugly head just before we depart for the base: he cannot attend the interviews because of some sudden UFO Museum crisis.

Running late as usual, we arrive at the airport. One of the interviewees is ready to bolt because he is tired of waiting for us. Great. It is so difficult to coordinate these things that you would think the participants would be a little more accommodating when we are in town to conduct business, but they usually aren't.

The interviews take all afternoon. First, Robert Slusher is interviewed at the former site of Bomb Pit #1, where on July 9, 1947, a wooden crate allegedly containing either three or four alien bodies was loaded into the bomb bay of a waiting B-29 (on which Slusher was a member of the flight crew) to fly them to Fort Worth, Texas. During the interview shoot, jet fighters suddenly decide to practice takeoffs and landings nearby. The sounds of the jet engines are deafening, then recede for a few minutes. This happens over and over again throughout the interview. I swear I can see one of the pilots smiling at us.

Ditto the interview of George Newling, which is conducted on the same spot. By now the jets have disappeared, but a little wooden vintage biplane has decided it is time to practice takeoffs and landings over and over again. Same result: starts and stoppages to the filming.

The last interview is with Robert Shirkey, a former flight operations officer at Roswell AAF in 1947. He has had to wait in the hot sun all afternoon while the other interviews were conducted, so he is not in a good mood. We all ride over to the old, still-standing flight operations building for his interview, which, to me, seems to take an inordinately long time to conduct; by the time it is over, everyone is exhausted. I have to get back to the UFO Museum to meet the dig volunteers and give them a short presentation. Shirkey, however, wants to be taken to dinner for his troubles. Fortunately, one of the MPH people volunteers to take Shirkey and his wife to dinner. We all breathe a sigh of relief.

7 P.M.

Back to the UFO Museum. I am shocked at how few volunteers have shown up. Turns out that at the last minute we lost about half of those who promised Don and me they would show. Nice touch. When the rubber hit the road, they bailed on us with one lame excuse after another. Some didn't even notify us; they just didn't show up. Really nice. Those who do show, however, ultimately work very, very hard throughout the entire course of the dig.

The head archeologist, Bill Doleman from the University of New Mexico, gives a remarkable lecture with visual aids on "becoming an overnight archeologist." Following short pep talks from Don and me, all adjourn to a fine dinner, compliments of the SCI FI Channel.

Donald R. Schmitt

Today the film and production crew begin their creative work. Learning that they had previously arranged to shoot all of the scheduled interviews at the UFO Museum, Tom and I argue that their procedure will reduce our witnesses to merely talking heads. After some reasonable, yet stubborn debate on our part, the producer and director concur; our military witnesses will have action shots out at the base, enabling them to demonstrate what they saw and where they saw it back in 1947.

The administrators at the old base, along with their security team, allow us to shoot right out on the tarmac. They have always been totally cooperative with us, though one never knows what might be coming in for a landing—B-52 bombers and F-117 Stealth fighters frequently use the extra-long runways.

Kate Leinster, independent UFO investigator

My journey begins at the Flagstaff Amtrak station in Arizona. I have a six-hour train journey to Albuquerque, New Mexico, where I am being met by Todd Fischer. After watching a beautiful sunrise (the first of many during this trip), I use my time to do some last-minute reading on Roswell, a mystery that has always intrigued me. I live in England, but am in America on a university exchange, so when my "e-mail buddy" Todd asked if I wanted to be a part of this historical event, I couldn't say no!

After what seems like forever traveling, I finally make it to Roswell and head straight for the UFO Museum, which as strange as it may sound, feels very surreal to actually be in. I meet most of the crew and other dig volunteers as we're given a crash course in archeology. Before leaving for a most welcome dinner, I, being a typical tourist, can't help but visit—and buy—half the gift shop!

The SCI FI Channel, in exchange for our hard manual labor, very kindly pays for our food and accommodation. Before falling exhaustedly into bed, I meet my roommate—a really nice woman named Debbie Ziegelmeyer who, along with her brother Chuck, educates me on all things UFO-related.

Debbie Ziegelmeyer

I pick up the rental SUV at the airport, meet my brother Chuck, and leave for Roswell, where a 6 P.M. meeting is scheduled at the UFO Museum. We knew the purpose—an archeological dig at the debris field made famous by Mack Brazel—but few other details were released to us ahead of time. We're just going on blind faith, and at our own expense.

We're surprised to see the SCI FI Channel crew here to document the event. I'm also relieved to learn that, along with SCI FI, a security team will be at the site. We were a little concerned about our safety; the 1947 Roswell Incident has always been associated with threats against individuals who ask too many questions.

We're given a brief summary of what's to take place over the next few days, a list of what we'll need (clothing, sunscreen, knee pads and gloves, etc.) and hotel accommodations courtesy of SCI FI. I'll be rooming with Kate, a very nice 20-year-old girl from England who attends school in Arizona. Chuck and I race to Home Depot and Wal-Mart for our supplies; then it's on to an 8:30 dinner—also courtesy of Larry and SCI FI—and another meeting.

Chuck Zukowski, independent UFO investigator

Debbie and I leave Colorado Springs around noon, and by 6 P.M. arrive at the predesignated location: the UFO Museum and Research Center in Roswell. Toward the back of the museum, we meet other individuals whose cryptic e-mail messages have also led them on this wild trek. It's at this meeting we also meet Dr. Bill "Indiana Jones" Doleman from the University of New Mexico, and learn that the SCI FI Channel is footing the bill for this event. I'm surprised to see SCI FI here, and wonder if this will turn into a media circus.

At this point, my sister and I think seriously about turning down the offer and just conducting alternate research while we're here. But after meeting with Larry Landsman from SCI FI and listening to Dr. Doleman and Don Schmitt, we decide to put our personal trust in these individuals and help as much as possible to achieve their goals (while also working toward our own).

Thomas P. Vitale, Sr. VP of Programming, SCI FI Channel

Running late, but I finally pack my bag, race from my apartment in New York City to Newark Airport, and make my United flight to Denver with about a minute to spare. Settling into my seat, I begin to unwind and think back on the events leading up to this trip.

When we first planned SCI FI's advocacy efforts with regards to Roswell and government secrecy involving UFOs, I was incredibly

Thomas P. Vitale (left), SCI FI's senior vice president of programming, and Larry Lands-man, SCI FI's director of special projects, look on as the dig continues. The SCI FI Channel sponsored this historic project as part of its ongoing public advocacy initiative calling for more scientific study of the UFO phenomenon. The channel is also working with Roswell investigators and the governor of New Mexico to have all "classified" documents on the Roswell Incident finally released.

excited. From these efforts came the idea that we would produce a spe-cial documenting our efforts in Roswell. Five years ago, in the summer of 1997, SCI FI produced a fifty-year anniversary special about the Roswell Incident, entitled *Roswell: Cover-ups and Close Encounters*. I helped supervise that special from my office in New York. This spe-cial was assigned to me directly, so I'm flying out to New Mexico to be on location for at least part of the shoot.

I'm really jazzed about going to Roswell, a place I've wanted to visit for years. While I was pleased with the 1997 special, a documentary about the history of the Roswell incident, this special would make history itself, as we're documenting the first-time dig of the debris field, as well as presenting new evidence to the case.

Getting to Roswell from New York takes longer than flying from New York to Rome. In Denver, I switch to a United flight to Albuquerque, with the final leg of my trip on Mesa Airlines (yeah, I've never heard of them, either . . .). I'm one of only five people to board a small prop plane—19 seats in all—which takes off about an hour after sunset for a 50-minute "milk run" flight to Roswell.

I don't particularly like flying, especially in prop planes, but this is the smoothest and most peaceful flights I've ever been on. (Go, Mesa!) The steady "white noise" from the plane's engine and propellers create a mood of silence. Outside my small window, I see the desert floor below us, faintly illuminated by the full moon that's just risen. It's a beautiful and somewhat haunting sight.

The plane touches down briefly at the Roswell, New Mexico airport; I'm the only one getting off here. The airport terminal is completely deserted, save for a lone security guard and Larry Landsman, who's picking me up. Larry is SCI FI's director of special projects, and one of the driving forces behind the channel's advocacy efforts. He's also assigned to the special and has been in Roswell a few days.

Larry tells me that the very spot I'm standing on had been part of the military base where the alien materials and bodies were allegedly taken to and flown from under a cloak of secrecy 55 years before. When the base closed down, it was turned into a public airport. Larry also informs me that they shot some footage here earlier today, with a man who worked at the base in 1947. This man had never before told his story publicly, and his story was compelling. My excitement is building.

We head into town, passing spots made famous by the events of 1947. These include the Ballard Funeral Home, which someone at the military base is said to have called, looking for small coffins. I'm told the funeral home doesn't look much different today than it did 55 years ago, at least on the outside.

We meet up with the production crew, volunteers and scientific team, who are just finishing dinner at Applebee's restaurant. I'm introduced to Roswell investigator Tom Carey and archeologist Bill Doleman, two gentlemen with whom I would spend quite a bit of time over the course of my stay. Over dessert, I learn from Dr. Doleman what a "contract archeologist" is, and about his interest in the incident at Roswell. Tom provides details of the case I hadn't known before, including how the location of the dig site was determined, and what new evidence had been uncovered since SCI FI's last special on the event.

After about an hour of meeting new people and having interesting conversation, I turn in, along with everyone else. Tomorrow's a big day, and since it's going to start at the crack of dawn, everyone wants to get as much rest as they can. I'm anxious to get started.

September 21, 2002

First day of "the dig"! Everyone met and formed a convoy of SUVs out to the crash site, located seventy-five miles to the northwest in Lincoln County. After more than a two-hour trip to the site (the final ten miles of which was over a two-track "road" of uncertain consistency), Bill Doleman presented the volunteers with an important souvenir and instant lecture on archeological field methods.

For many, the dig was a ten-year-old dream come true; a chance to prove, once and for all, that a UFO, not a balloon, met its fate in this sheep pasture back in 1947. To the volunteers, it was an interesting, exciting opportunity to help find the answer to a burning question, and perhaps be part of a history-making event. The archeologists saw it as another contract job, albeit one that was a little different from what they were used to. The usual "archeological stuff"—soil stratigraphy, erosion and deposition patterns, and so on—would of course be useful, but the chance to solve a legitimate twentieth-century mystery appeared to really interest them. The SCI FI Channel, which was funding the entire cost of the expedition, hoped to make history, and employed MPH Entertainment (the production company responsible for *My Big Fat Greek Wedding* and *The Lost Dinosaurs of Egypt*) and a still photographer from Santa Fe, California, to record everything for posterity.

As digging commenced in the prearranged test pits near the alleged gouge using approved archeological field techniques, interviews on loca-

At the crack of dawn on September 21, 2002, the caravan of archeologists, scientists, and representatives from the SCI FI Channel head out to the debris field, or "skip site," as it is called, seventy-five miles outside Roswell to begin their grueling work. About the first sixty miles of the drive is on highway, while the rest is along dirt roads. The drive takes about two hours.

tion for the SCI FI documentary about the dig included a remarkable impromptu conversation with one volunteer whose late father had factored heavily in the cleanup of the very site she was digging. Not all of the volunteers were comfortable with the way things went, however. The digging was hard work, nothing of significance had yet been found, and there seemed to be too much emphasis placed upon the procedures rather than the object of those procedures—finding "HMUOs" ("Historic Material of Uncertain Origin"), as the archeologists refer to any manufactured artifacts that couldn't be readily identified. A Native American projectile point would not be considered a HMUO because it could be identified as such, though a balloon or tinfoil fragment would be labeled a HMUO until it was positively identified, since it did not belong there. A piece of interplanetary spaceship would certainly qualify as a HMUO, and it was this type of artifact—what some call the "Holy Grail of Roswell"—that everyone dreamed of finding.

Dr. Bill Doleman

We had all agreed to meet at 6:30 A.M. By 7:30 we're all ready, and a caravan—including a 25-foot RV, the MPH production team, some volunteers in their own vehicles, more volunteers in the OCA van I'm driving, the SCI FI folk, and Don and Tom (the project's "technical advisers")—head out of Roswell to the skip site.

Once at the site, Bob and Lou continue setting up grids while I get "miked up" and give the volunteers their second "instant archeologist" lecture. After delving deeper into the heart of the excavation process— i.e., exactly how to move dirt from the ground and into a screen for sifting, not to mention the all-important paperwork—I give each of them a real Marshalltown trowel, to use and keep as a souvenir. The Marshalltown trowel is to the field archeologist what the six-gun was to the cowboys of yesteryear (though I guess "lariat" would be a more realistic analogy). It's clear to me that the volunteers' enthusiasm will more than compensate for any lack of experience. Meanwhile, I'm starting to get used to having a large, expensive video camera hovering near me whenever I talk.

Next we review the site integrity measures I had designed to keep anyone from "salting" the site. These include taking photos of the undisturbed surface of each grid prior to excavation, as well as of the bottom of any unfinished grids at the end of the day, after which they're sealed with plastic and dirt. We then repeat the process first thing the next morning.

After reviewing the methods and answering questions, the crews of two—Lane and husband Jerry, Nancy and husband John, Chuck and sister Debbie, Howard and Kate—set to work moving dirt under the supervision of us OCA archeologists. The testing has begun!

We have one crew excavating in each of the two electromagnetic conductivity (EMC) anomalies Dave had discovered along the "furrow baseline"—one doing the "baseline stratigraphy" pit in the middle of the site, and one on a test pit in an area that had obviously been burrowed (or "bioturbated") in the past. Bob and Lou cruise around the rest of the afternoon answering questions, most of which revolve around figuring upper and lower elevations for each excavation level.

William Doleman distributes Marshalltown trowels to all the volunteers and gives them their second "instant archeologist" lecture. He is quick to point out the inherent dangers of the desert, such as tarantulas, rattlesnakes, and centipedes.

Given that we had eight volunteers instead of the planned 12 or more, we had already decided to eschew five-centimeter levels in favor of the more efficient, if lower resolution, 10-centimeter ones. When Lou, Bob and I had shot in the location of the test pits, we also used the transit to read elevations expressed in terms of the arbitrary grid system we established the first three days. This would provide the excavators with "starting elevations," from which they could compute the beginning and end elevations of each excavated level.

Their measurements involved using a string and contractor's line-level tied to a large nail in the ground (whose elevation was shot by us), plus a metric measuring tape to determine the depth of the excavations. The concepts and arithmetic are simple, but I've never worked with a new excavator (myself included) who hasn't taken an afternoon

From left to right: Investigators Don Schmitt, William Doleman, and Tom Carey examine a potential site for one of the test pits. The location of individual archeological test units are based on: (1) results of the aerial photographic inspection; (2) locations of subsurface anomalies as revealed by the EMC study; and (3) available anecdotal accounts of the locations of the impact and the resulting debris field.

to become comfortable with the methodology. Our volunteers have it, plus the rest of the unusually complex project-specific methods, down pat by day's end.

I spend a good part of the remaining day confabbing with the folks from SCI FI and MPH Entertainment, as well as discussing the project with Bureau of Land Management–Roswell archeologist Pat Flanary and his coworkers. Pat seems impressed with our overall strategies (which were detailed in the testing plan filed with and approved by the local and state BLM offices) and their rigorous implementation in the field. He's also impressed by the speed with which the volunteers have picked up a pretty complex field methodology. It's clear to me that they're all quite intelligent, on top of their interest in the project.

One important conversation involves me, Don, Tom and Larry Landsman of SCI FI. I said, "I know that when I brought this idea up before, everybody said it wouldn't be a possibility, but if you want to find evidence of the furrow and do it efficiently, a backhoe is the only way to go. If it's there, the backhoe will find it, and I know just the guy for the job. His name is Eligio Aragon, but everyone calls him 'Alley' because his business is called Alley Cat Excavating. I've worked with him for over 10 years, and he's been doing archeological backhoe work for over 20 years. One professor even flew him to Egypt to do his excavating for him." We agree to ask the BLM archeologist, who I call later; he doesn't see why we can't use the backhoe, but he'll be out tomorrow.

The sun's setting by the time we leave for the long drive to Roswell. Before doing so, I take the "end-of-day" record photos of each test, and the volunteers seal them. Like today, the ensuing three days would be 12 to 14 hours long, long drives included, but we're all excited and having fun.

Back in Roswell, SCI FI buys us a great dinner at a local eatery, though by dinner's end I just want to climb in the sack. I call Alley first, only to find out he's in California, serving on somebody's pit crew at the Laguna Seca race! His wife says she'll let him know. I set my wake-up call for 5:30 A.M.—which will come way too soon—and hit the hay.

Thomas J. Carey
6 A.M.

The first day of the actual dig. Everybody meets at a predetermined location and forms a convoy to head out to the Foster Ranch debris-field site. It takes about three hours to get there, and we are delayed in starting when a few vehicles—including that of the head archeologist—have to fill up their gas tanks. I had filled ours the night before, to avoid inconveniencing others. Oh, well.

I am interviewed at the site for the documentary by Melissa Jo Peltier, the "P" in MPH Entertainment. The interview goes well, with the only problem being which hat to wear during it. I wind up wearing one of the sound men's "Save the Whales" baseball cap, but I almost wear a pith helmet (as suggested by Bill Parks). My wife would laugh me out of the house if she saw me wearing that; it reminds me of something an archeologist would wear in one of those old "Mummy" movies.

The interview is cut short for lunch, but I am told that it will be resumed later. It isn't.

Volunteer archeologist Nancy Easley Johnston excavating her test pit. It was Nancy's father, Provost Marshal Major Edwin D. Easley, who was in charge of securing the crash site in 1947 and cleaning the area. Fifty-five years later, Nancy searches for debris that her father's team may have left behind. On his deathbed, Major Easley used the word "creatures" when referring to the victims of the crash.

I talk to as many of the volunteers as I can, to thank them for their efforts and especially for showing up. I meet Nancy Easley Johnston, her husband John, and Dr. Harold Granek for the first time; they have come from Texas for the dig. Nancy's father, Edwin Easley, had been the provost marshal at Roswell AAF in July 1947, and was in charge of security during cleanup operations at the very site we are now standing on. Our hope is that Nancy, who was present when her father acknowledged to her family in his last days that bodies had been recovered from the crashed UFO, will consent to an interview. She does, expressing how she was making a connection with her deceased father. Besides revealing for the first time his role in the 1947 events at Roswell, he explained to his family how he had promised the president [Truman] that he would keep the secret. According to Nancy, when a family member showed her father one of the books about Roswell, he turned his head away and said, "Oh, the creatures."

Don has to go back to Wisconsin on a family matter.

Donald R. Schmitt

I rise at 3 A.M. and get ready to return to Wisconsin. I had promised a year before that I would attend my goddaughter's wedding, but why does it have to be this very weekend? The one event we've been working toward for over a dozen years, and I have to cut out before it starts! But the happy expression and hug from Heidi once again demonstrates to me that family comes first. I just have to hope that Tom is not walking into any more cactuses back in New Mexico.

Kate Leinster

Get up before dawn in order to get to the dig site, some three hours away; it's a journey that over the next few days I will get to know well! Everyone meets in a Wal-Mart car park that's nicely decorated with a big alien; am I in Roswell?!

I will never forget arriving at the dig site for the first time. It is amazing how the location can be pinpointed so precisely; the whole landscape is so barren, apart from two portable toilets glistening in the sunlight like beacons! It feels both strange and exciting to be standing in such a controversial and historical place.

Bill Doleman, the head archeologist on the dig, gives us novices another lesson on the basics of archeology. I am quite shocked by how few volunteers there actually are, given the great interest that surrounds Roswell. I put it down to the fact that the whole operation has been quite hush-hush; I wouldn't be digging around in the dirt if Todd hadn't told me about the project.

The volunteers are paired up. My partner is Harold, a nice guy from Texas. Everything takes a little getting used to, but we soon have

Aerial shot of some test pits. The volunteer excavators, their supervisors, and the project director all kept detailed notes on test unit stratigraphy and any recovered artifacts using standard grid excavation forms. All units were then given a unique study unit number. In addition, plans and profiles of all units were photographed—and drawn as well, when deemed necessary.

our own little production line going—Harold digs and carries the heavy buckets while I, the "weak female," sift the mounds of dirt. While searching desperately for any sign of what Bill calls an "HMUO"—"Historic Material of Uncertain Origin"—not only am I rather burned by the end of the day despite my trusty Factor 30 sunscreen (I am from England; I'm not used to the sun!), I'm two shades darker with all of the dirt I've managed to sift on myself.

After a hot, sticky, unusual day, we are treated to a lovely meal, followed by an even lovelier—and most welcome—bed.

Debbie Ziegelmeyer

We meet at the Wal-Mart parking lot in Roswell around 6:30 A.M.—SCI FI crew, archeologists, Don Schmitt and volunteers—and are caravanned to the site. We arrive around 8:30 A.M., after passing

Controlled excavations are conducted under the supervision of William Doleman and his two archeological assistants. When any human-made archeological features are encountered, they are left intact for future investigations, and the test unit is adjusted accordingly. Excavation is by hand, and all excavated fill is screened using one-eighth-inch mesh.

through security checkpoints. Bill Doleman presents us with trowels that we can keep, and a crash (no pun intended) course in archeological excavating.

After lunch, we finally start digging our holes. The work is hard, but I'm excited to be a part of this monumental event. At the end of the work day, we all pose for a group picture. I don't ever remember being so dusty and dirty; by the time we get back to Roswell, we barely have time to wipe down and change clothes before dinner.

The restaurants in Roswell close around 9 P.M. Larry Landsman is able to convince one to stay open for our group each night so we can have a good, hot meal. Unfortunately, we have to wait to shower, so by the time we eat, have a meeting, make our way back to our hotel and wash up, it's 1 A.M. before we get to bed.

Chuck Zukowski

Saturday morning starts with our 6:30 A.M. rendezvous at Wal-Mart, watching the sunrise while sipping our gas station 30-weight coffee. Larry Landsman barks out orders in the distance, so we gather our gear, hop into white unmarked SUVs and motor to the Mack Brazel debris site. We're in constant communication with handheld radios as we follow New Mexico's finest archeologist, Dr. Bill, to look for the un-lookable.

We arrive at the debris field within a couple of hours, and with a motor home chuggin' in as a command post, we put on our student caps and begin an intensive course in archeology. Oh, did I forget to mention the armored vehicle and all the armed guards surrounding the area? Oh well, no big deal; I had done archeology work in the past, and my sister had been on underwater treasure hunts in the Keys, so we were used to the armed guards. (Not.)

We're ready to take anything Dr. Bill can throw at us, or so we think. Then we start hearing commands, as if Patton had returned from Europe. "We will start by digging a one-meter by one-meter hole, 10 centimeters in-depth at a time, at the pre-designated locations," Dr. Bill says. "We will take dirt samples at every 10-centimeter level, only after it's been properly sifted. You will dig at least six to eight levels deep, and you will enjoy it! You will also make sure all paperwork is properly done when excavating your site! If not, I will personally come over and bonk you on the head!"

Three hours later, and our brains are ready to collapse with Dr. Bill's brilliant information. After lunch, with our spiffy grade-"A" archeologist spades in hand, we start clearing dirt—"digging a hole," for all you laymen out there—at our one-meter by one-meter sites.

Now, clearing dirt is not as easy as it seems. First, you are constrained by a one-meter by one-meter that's roped off and held in by four spikes, one located in each corner. Then you have a level spike in one corner, which has a string attached a certain distance in centimeters from the surface of the ground. The archeological surveying equipment determines the height of the string on the level spike. This gives you a precise starting point to measure the depth of your dig through each level.

Work on each level consists of taking your trusty trough, placing it at the desired angle to the ground, then cutting away dirt a few centimeters at a time. All the while you try not to destroy anything that could be underground, or jam your fingers into the hard-packed surface. (Of course, I didn't do that.) Each level is 10 centimeters deep, and must be properly measured and maintained to have a proper dig.

To properly measure the depth of your dig for each level, you must use the leveling string attached to the leveling spike. Connected to the leveling string is a bubble level. You have to stretch the leveling string across your dig while using a ruler to measure the depth. The string and ruler create a "T" formation, with the top of the "T" at the bubble level.

Moving the ruler across the bottom of your dig while the string is positioned at a 10-centimeter increment, plus keeping the bubble in the center of your level, ensures the floor of your dig is exactly level to the surface of the ground at a 10-centimeter depth. Now you must repeat this measuring system for every 10-centimeter level you dig. (Whew! Confused? I was, at first.)

Okay, now sifting the dirt. All the dirt taken from each level of the dig must be sifted properly using an ACME 2002 sifting screen. It's roughly a four-foot by four-foot square box made of two-by-fours, with a wire mesh screen attached for the base. The sifting screen is then attached by rope to the inside center of a large tripod, allowing it to sway back and forth. Standing firm and holding the two-by-fours of the screen, you violently shake the screen while yelling colorful metaphors (you don't really need to yell, but it helps when you get a splinter in your hand). The amount of dirt we normally sift fills a five-gallon bucket—a lot of dirt, and a lot of dust.

Notes:
1. Sifting dirt should be done with your mouth closed. (If you could close your nostrils, that would help, too!)
2. Sifting dirt next to someone trying to dig their site could cause bodily injuries to the person sifting.
3. Wearing shorts while sifting causes the hair on your legs to attract the sifted dirt like a magnet, resulting in Sasquatch-looking legs. Which isn't too bad unless you're female and forgot to shave that day.

Why do you sift? Sifting gives you an opportunity to look for artifacts. An artifact—or "HMUO," as Dr. Bill refers to them (it stands for "Historic Material of Uncertain Origin")—is any tiny object that looks out of place within the environment you are excavating. These objects

Volunteer archeologists Chuck Zukowski (left) and Debbie Ziegelmeyer log in and tape up a sample of soil from their excavated pit. All Native American artifacts encountered on the surface or in test excavations are placed in envelopes or bags with labels showing the excavation unit's provenience information and catalog number. Sediment samples taken from test units are sealed with high-quality tape to ensure their integrity during and after transport.

could be a piece of pottery, an arrowhead or maybe a piece of space-ship. Simply put, if it doesn't seem to belong there, it's an artifact. If you find one, you bag, log and transfer it into safekeeping. In our case, "safekeeping" is an armored vehicle surrounded by armed guards.

The artifacts we're looking for are anything that resembles eye-witness accounts of the debris seen back in 1947. Of course, if we find any balloon fragments, we've been told not to say anything and hide them down our shirts. Seriously, we sift to look for any artifact that could prove or disprove the veracity of this excavation site.

The end of the day yields a hole 40 centimeters deep, which is well documented by the SCI FI photographers and cameramen. And even though I stop from time to time to discuss the secrets of the universe with them, we achieve a great deal of progress. We need to finish this particular hole by Sunday morning so we can start another after lunch.

You can't just walk away from an unfinished site for the night. You need to get out your trusty camelhair brush and dustpan and sweep out your site until the bottom is as clean as the proverbial whistle. At that point, one of the archeologists ventures over and takes a picture of the inside of your dig, to ensure it's not "salted," or tampered with, while you're away. Once photos are taken, you cover the hole with a tarp and secure it for the evening. Then you secure your buckets and shovels and lower the sifting tripods. You do this because it's been known for cows to come through your site and try to finish what you started. Then they get all the credit. Stupid cows.

We head back to Roswell for a very late dinner, where we discuss the day's events and make war plans for tomorrow. Then it's time to crash.

Thomas P. Vitale

I wake up at 5:30 A.M. and am downstairs for breakfast by six. It's a beautiful morning in New Mexico—mostly sunny, with a crispness in the air. As a New Yorker, I always notice the clean air when I travel to wide-open country.

Larry Landsman, Tom Carey and I load up into our rented SUV and head to the Roswell Wal-Mart for a 6:30 meet with the volunteers, production team and scientists. I see our producer, Melissa Jo Peltier, who skipped last night's dinner in favor of sleep. I also notice that even the Wal-Mart has aliens painted on its windows; I assume it's for the benefit of tourists.

After a bit too much waiting, our caravan of six SUVs and a production trailer head out to the debris field, 75 miles outside Roswell. The first 60 miles or so of the drive is on the highway, while the rest is along dirt roads. The drive provides me a rare opportunity to spend two hours in a car with Tom, and to ask him detailed questions about his work and the various theories regarding what happened at Roswell. I'm sure Tom had heard most of my questions from many, many others before. He's passionate about getting the details of what he knows about the incident out to the public; he thinks that if people knew what he knows and kept an open mind, they wouldn't be skeptical at all. It's a fascinating conversation; I learn a great deal more than I already knew about the timeline of events, and about various pieces of evidence.

The leading theory from Tom and his research partner, Don Schmitt (who'll rejoin us tomorrow), is that there were three crash sites. First, there's the debris field—sometimes called the "skip site"—on the Foster Ranch, where we were heading. Witnesses claimed to have seen something crash down on a field in the middle of the night. Tom explains that when all of the evidence and eyewitness and secondhand accounts are pieced together, the picture becomes pretty clear. The craft crashed in this field, creating a furrow in the land and leaving a lot of debris behind. It was able to become airborne again, but was severely damaged. About two and a half miles later, bodies might have fallen from the damaged craft. The craft continued to fly, probably in an out-of-control fashion, and then finally crashed after flying about 15 miles or more closer to Roswell. Some believe that bodies were recovered at this second site as well. Tom and Don searched for this second site, and believe they might have found it, or at least come close.

As we near the debris field, we first check out the actual cabin where intelligence officer Jesse Marcel and ranch foreman Mack Brazel stopped for the night before proceeding to the field. We take some footage and learn from Tom the significance of the cabin, which really brings home to me how much the world has changed in just 55 years. Back then, travel was much slower. Marcel, a top military officer, and Brazel ate beans out of a can and slept on the cabin's hard wood floor, waiting for first light. Our trip from Roswell to the debris field took less than two hours on the Interstate. And there are motels everywhere, as well as comfortable SUVs and campers. Back then, communication between the debris field and the city was nonexistent. Today, those of us at the site who aren't equipped with satellite cell phones complained because we can't get a signal.

We soon approach the field that our research and scientific investigation tells us is the skip site. The field unfolds as part of an enormous scrub desert sitting in the middle of a huge ranch, on government land managed by the Bureau of Land Management (BLM). Many people find the desert to be lonely and desolate, but there's a special beauty to it as well; the land features hardscrabble sandy dirt, low cacti and succulent plants, small weeds and rocks. The field spreads out in front of us for a few hundred yards to a low ridge—possibly the one the ship is reported to have hit. To our right, the land stretches in an unchanging vista for as far as the eye can see. To our left, the vegetation grows a bit denser, the cacti progressively taller, as mountains rise up about a mile in the distance.

Before I left for Roswell, it surprised me how many of my friends and associates confused the Roswell incident with stories of Area 51, which is in Nevada, not New Mexico. Many people just don't know how important this area is in terms of aviation and military history. One of the volunteers gives me a sense of the area's geography, pointing out the direction of the various crash sites and Roswell itself in relation to where we are, and identifying the various mountain ranges. I'm told that the *Enola Gay* left from Roswell with its powerful payload in 1945, and that this debris field is about 200 miles from White Sands, where missile technology was developed and tested in those days.

Walking onto this unchanging field, I truly feel like I'm taking a step back in history—not just to July 1947 and the time of the Roswell incident, but also into my personal family history. Back in the mid-fifties, my father was a photographer with the Army Pictorial Center in New York. He was sent on assignment around the country with his unit, and spent about six months out in White Sands filming missile launches. At night, while lying on the sand, waiting for missile launches to film, he'd look up into the night sky and see all sorts of moving lights in the star-blanketed sky. Some were shooting stars, some were missiles or military aircraft. But others . . . he and his buddies couldn't identify. I know that, with all that the military did in New Mexico in the forties and fifties, if we are being visited by aliens, this area of the planet would be ripe for their surveillance.

Once at the site, Bill Doleman takes over. He warns everyone about the inherent hazards of the desert. Those of us who are city folk listen intently; we had already seen a tarantula on the dirt road as we drove in. We're told to watch for rattlesnakes and centipedes, both of which are dangerously poisonous. (I also hear that the largest centipedes are probably deadlier than most of the snakes we might find.) Somebody kicks one of the big, flat gray patches sitting on the dirt, revealing a very large centipede. We learn that they prefer to stay under these "sun-baked cow pies" to keep out of the sun and gather moisture. We're also cautioned to stay away from the cacti, not only because of barbs (obvious even to city folk!), but because centipedes and snakes might be hanging out beneath them, as well. The same goes for loose rocks.

After the much-needed wilderness refresher course, Bill starts to explain the science of what we're looking for here. Except for grazing animals and small burrowing critters, the field is basically untouched land; in particular, it has been undisturbed by humans. If something did crash here in 1947, two things were of crucial importance. First, there is no guarantee that the army would have been able to completely clean the field. A small burrowing animal could have grabbed a piece of wreckage and buried it in its hole. Over time, the object could fall even deeper, in a process called "bioturbation." The key point is that by excavating the field, we might be able to find a piece of that debris.

The second point Bill makes is that if a large object did crash in the field, then the ground would show signs of being "different" than the ground nearby. Remember, these fields aren't farmed and haven't been built on, so the ground below the surface has been untouched (except by animals) for thousands and thousands of years. Using special equipment, Bill and his team found a strip of ground in the field that was disturbed in a major way in the relatively recent past. This strip had greater conductivity than the land around it, which means it had looser soil and a higher moisture content. That led the scientists to believe that something indeed could have crashed here. This strip of land, the possible crash gouge, had been marked off by little surveyor flags a day or two earlier.

Next comes the scientific fieldwork. Bill and his team explain to the volunteers the process by which archeologists work with and dig on historical sites, whether they are looking for dinosaur bones, artifacts from ancient cultures or evidence of alien visitation. The work is painstakingly tedious, but also incredibly interesting and exciting. The first key to a dig is to keep the site "clean," lest the work be questioned (i.e., you don't want to be accused of putting items into the site and claiming that you actually took them out). The second is to dig slowly and carefully so that you don't miss anything or damage whatever you find, and that you can identify at what level of depth your artifact was buried.

After helping the volunteers set up little squares for digging, 50 centimeters to a side, the scientists teach them how to dig in a very slow and meticulous fashion with small hand shovels. They're shown how to take soil, bucket by bucket, from each layer of depth and sift it through screens. Any "Historic Material of Unknown Origin" ("HMUO") is to be put into paper bags, which are then sealed and written on with details about where they were found. Soil samples from different layers of digging will also be bagged, and Bill says the same goes for any Native American artifacts discovered. These articles will be studied before being reburied in the field, as federal law mandates. With his first shovel-full, Bill actually finds a Native American artifact—not an arrowhead or anything so obvious, but a sharpened rock that would have been used as a tool hundreds or thousands of years ago.

All the while that Bill teaches, the production crew rolls. Larry and I are careful to stay out of camera range.

Before I left for Roswell, some people in the SCI FI office told me they expected the dig to be big and fast, with a lot of guys with big shovels—almost like a construction site. The important thing for me to note here is that this dig is being conducted slowly, with the utmost scientific care and methodology.

The weather is perfect—low- to mid-60s and mostly sunny. Just warm enough to be comfortable, but not too hot to slow down the diggers and scientists. Around midday, a man from the Bureau of Land Management shows up. He supposedly knew the person who helped get us the permits for the dig, and he came out "to see how everything was going." He's a nice guy and seems earnest, but many of us are suspicious about his intentions. After all, it is a Saturday, and we're a two-hour drive from town, so he clearly wasn't "just in the neighborhood and decided to drop by." We wonder if he's here to shut us down and help the government continue to hide the truth. Perhaps he's here to report back to "someone" on our progress. Or maybe we've all seen too many episodes of *The X-Files,* and he really is here just because he heard something interesting was going on and was curious.

Another thing worth noting about this man's visit was that he put forth a theory of what he thought had happened at Roswell—that the military had been testing some new technology that possibly got some people killed. The tests were classified, since it was the Cold War, and the truth was buried. His explanation is very close to the official government explanation, enhanced by his own personal speculation. Afterward, Tom Carey easily shoots holes in his theories, so again we wonder whether "Mr. BLM" was sharing his own thoughts or following someone else's agenda. Our speculation about our government visitor is fun, but the truth is that he was nothing but friendly, smart and courteous, and he seemed truly interested in what we were doing.

We later have a second visitor—the current ranch foreman, who's introduced to me as today's version of Mack Brazel. We ask him if he has ever seen a UFO. He says he hasn't.

Except for a lunch break, no one stops working until a little before sundown. Bill starts calling the volunteers true "junior archeologists" by the end of a day that's filled with filming, teaching, digging, interviewing, exploring, photo-taking and an overall feeling of being part of history. Besides fascinating interviews with Bill and Tom for our SCI FI special, the best interview comes from Nancy, a volunteer whose father played a crucial role in the Roswell incident in 1947; in fact, she joined the dig in honor of her father's legacy. Her interview is incredibly compelling, though I don't want to give away details here.

While working, Larry and I take time for a little fun stuff. We take lots of personal photos of the day, and even collect some soil and rocks as mementos of the dig. Once everyone finishes digging, we debrief and head back to Roswell, stopping only to film some beautiful sunset moments. The group enjoys a delicious dinner at a Mexican restaurant in town, then turns in to get ready for tomorrow.

September 22, 2002

Dig Day Two. The weather had improved, and the volunteers were now experts in archeological field techniques. The MPH Entertainment crew was in Roswell, which meant no distracting interviews. The RV and two Porto-Potties were in place at either end of the site. As Bill Doleman stated, "The archeology gods were looking after us."

In addition to bagging any HMUOs recovered, soil samples from each test pit were screened and bagged for later analysis of the possible presence of microscopic fragments from whatever crashed there in 1947. Digging became increasingly harder as the volunteers dug deeper. By late morning a crisis brewed among frustrated and angry volunteers who felt they weren't digging in the optimal locations to find HMUOs. Meanwhile, geophysical prospector Dave Hyndman conducted the second phase of his subterranean studies via a very sophisticated, high-sensitivity metal detection search, concentrating only on areas identified by the archeologists as possible locations where running water would have likely deposited waterborne artifacts if such were present.

Dr. Bill Doleman

This morning we're much more efficient at getting out of town in that we don't have to wait for the RV, which is now stationed at the site. Entering the project area, we're met and checked by guards from Fortress Security. They guarded the site last night, and will continue to do so until project's end.

Sunday, like yesterday, is slightly overcast, but after all the nasty wind and cold we had mapping prior to the volunteers' arrival, the weather's great—cool but not cold, clear but not glaring skies, and no breezes approaching "obnoxious" or "impossible." The archeology gods are looking after us.

Everybody goes right to work as soon as I take the "start-of-day" record shots. Everything proceeds smoothly; the MPH folks are in Roswell doing interviews, so there are no distractions. Bob, Lou and I return to mapping until the volunteers start asking how deep they have to go. I explain that these are our first explorations, and that until we have a better handle on the site's stratigraphy, both natural and any created by a 55-year-old impact (if present), we need to keep digging. "Stratigraphy" refers to the dirt layers, their characteristics, and what those characteristics have to say about how they got there and how old they are. Some are so-called soil characteristics that result from the action of natural processes on the dirt after it's deposited.

Dirt is flying now that they have the system down. Soil samples from each level are collected from under the screens for possible microscopic and chemical analysis later; any non-natural things found that aren't readily identifiable as a Native American artifact or historic artifact like a nail or glass fragment are bagged as an "HMUO"—"Historic Material of Uncertain Origin." This term was coined to handle anything that might be the "debris" we're looking for.

The search is guided by the geophysical prospection explorations and any anomalies detected, the locations of areas of apparent nonrecent animal burrowing, inspection of the site's topography and determination of low-lying areas where any remaining debris might have been washed down and buried. Almost all of the dirt and deposits at the site were the result of eons of weathering of the underlying bedrock into fine-grained sand and dust, and its subsequent transport by surface-flowing water during big summer thunderstorms. That said, we also dug in low places!

Dave Hyndman shows up to conduct the high-sensitivity and high-resolution metal detection survey with an EM-61 device. Since he isn't budgeted to survey the entire site, we discuss where his time will best be spent and settle on several areas, including the low-lying areas where "stuff" might be buried. He and Sid get to work, spending the rest of the afternoon at it before heading home to process data. Dave says he'll have the results later tonight.

Meanwhile, the volunteers begin to get deep enough that I can make some initial assessments of the stratigraphy in various parts of the site they're working on. I think I see evidence that they're into dirt that's way older than 1947. We stop mapping and shoot in new grids for them, targeting those low-lying areas where debris might have collected during 55 years of storm-generated "slopewash," as the washing-down-and-burying process is called.

The volunteers spend the rest of the afternoon in their new grids, and we get some more mapping done. At the end of the day, we transfer all soil samples and artifact bags—HMUOs included—to a locked security van brought to the site by Fortress Security. We decided that would be preferable to us taking stuff back to Roswell every day. FSI personnel stay on site all night long.

Same long ride back to Roswell, and same great dinner with volunteers and folk from MPH and SCI FI. Back at the hotel, I call Dave, who says he has a "very exciting anomaly" for us to investigate, and he'll be out tomorrow to show me his results. Wow! If I wasn't so tired I might have trouble sleeping.

Thomas J. Carey
9 A.M.

I drive to a McDonald's at the south end of town to interview 92-year-old Floyd Green; I am told that he joins a breakfast klatch of old-timers there most mornings. Floyd is indeed there, and we find a separate table for the interview. He knows all of the Roswell players from 1947, including Mack Brazel. "I wouldn't have believed this except for old Mack," he says. "If he said something was true, you could take it to the bank. That's why I believe that there was something to this." I take some pictures of Floyd, thank him and head back to the UFO Museum.

Don is still back in Wisconsin on a family matter, so for the first time I head out to the dig site on my own. Small flags have been placed all along the route to the site. Even though I have been to the site at least 10 times, I still wouldn't find it without those flags. I can't imagine how Don, with one good eye, and Todd are able to find it at night.

I am exhausted when I get back to Roswell, so I stop at Arby's for a quick roast beef sandwich. I retire early, but the phone rings; it's Larry Landsman, who wants me to join everyone else at The Cattle Baron. I reluctantly accede to his wishes.

Don returns from Wisconsin and joins us at the restaurant.

Kate Leinster

Up early once again. I call my mom back in England, though I don't feel like I can say too much; what with all the nondisclosure forms we had to sign, I don't want to risk being sued! Besides, you never know who is listening to your phone conversations. (OK, so maybe I have seen one too many episodes of *The X-Files*.)

At the site, I am starting to ache, but still feel buzzed to be here. After much digging, Harold and I think we have come across an HMUO. We're pretty excited. It looks like some kind of rubber; could it be alien? Actually, no. It turns out to be the sole of Harold's rapidly disintegrating shoe! However, I do find something interesting later; some very thin "plastic." Taking no chances, I bag it up.

Arrive back late in Roswell. I miss dinner because I really need to shower. I've never felt so dirty in all my life.

Debbie Ziegelmeyer

The day starts with us promptly meeting at Wal-Mart again around 6:30 A.M. I'm tired; we all had a late night. There isn't time for breakfast, so we get a quick coffee and box of donuts to go. We drive to the site in our own rental SUV, which is an exact match to the vehicles SCI FI had rented.

Today the SCI FI crew has interviews to do in Roswell, so we have a little more time to work, and no one to interrupt us. After a hard morning of digging, we break for lunch and another meeting. Bill Doleman explains the reasoning behind the size, depth and placement of the holes we've been digging. We walk the area and take some pictures while Bill, Bob and Luey find us new dig sites that have better prospects for UFOs, or at least pieces of one. After all, isn't that what we're looking for? Then it's back to more digging, followed by another long ride back to Roswell, a late-night dinner, a shower and bed by 1 A.M.

Chuck Zukowski

Today starts as usual—in front of Wal-Mart by 6:30 A.M. This time, however, the SCI FI crew stays in town to tape interviews. After a cup of Joe and gas for the vehicles, we make pretty good time getting to the dig; Dr. Bill likes to drive fast.

Upon arrival we all head to our dig sites and start removing the tarps. Before we can proceed with the digging, one of the archeologists has to come by and take another picture of our hole; ensuring that the picture taken the previous night matches the one taken the next day keeps the site authentic, and not subject to a salting controversy.

Our particular site becomes tough to dig; Debbie and I are hitting hard-packed dirt at 50 centimeters. It takes over an hour to dig another 10 centimeters, and by the time we hit 70—around 11 A.M.—we land on a white rock layer.

We volunteer diggers start to get frustrated. As a researcher, I know debris will not be found in 2,000-year-old dirt. I decide to challenge Dr. Bill, who explains that my particular site is a "sedimentary site," dug so archeologists can have an idea how the layers of soil are laid out. Seventy centimeters is good enough, and now the archeologists can use our site to monitor the different soil layers with underground moisture seepage.

Before Debbie and I start our next hole, lunch proves somewhat interesting; it seems we aren't the only ones frustrated with digging down 2,000 years. Other volunteers are starting to feel the pain of hard-packed dirt with rocks. You have to understand—we're using garden spades, not picks or shovels. Crunched knuckles plus bent fingers equals anger and frustration, so we all gather for a discussion in the motor home during lunch.

After we eat our gas station grub, Dr. Bill uses writing utensils to explain why we're doing what we're doing. He illustrates how important it is to know the soil levels at different parts of the debris field, and how very important it is to know the contour of dirt. This is archeology at its finest; still, we're here to excavate for artifacts, not dig to China, so I decide to say something again. Why? Because I'm stupid! But as one of only two actual UFO researchers here, I step forward.

"Bill," I say, "please draw me a sloped line. Now slope the line down to a straight line to represent a hill. Now, if an object skipped off this area here at 'X' amount of speed, scraped the surface of the ground at 'X' amount of distance here, then over 55 years of erosion, where would be the best place to dig?"

Bill thinks about it for a moment and begins to realize where we're coming from, and for a brief minute there's some bonding in the air. We begin to understand archeology work, and the archeologists start to understand the many personal reasons we've devoted our time to this project. There's a glimpse of sun and a sliver of a rainbow in the air as we all start seeing eye-to-eye.

After lunch we take a break while Dr. Bill and his team check their surveying equipment and focus on where the next digging sites will be. We're excited, and within an hour we're working in new excavation sites, each carefully documented and carefully orchestrated. Dr. Bill is also hard at work; this guy never stops moving. I hold him and his team in the highest regard; such professionalism in today's world is hard not to notice. This is archeology Indiana Jones would be proud of.

The afternoon yields some HMUOs, and they're bagged, documented and secured in the armored vehicle. Everyone's busy at work; the muted sounds of humming in the background reminds me of the sounds the guards made at the witch's castle in *The Wizard of Oz*. Meanwhile, our wizard from the Emerald University hides behind not a curtain, but a clipboard, planning our next course of action.

Like we did yesterday, we follow the proper procedures to secure our sites when it's time to leave. With muscles aching, sunburns pounding and dust in our eyes, we once again say goodbye to another day in paradise. Paradise only a passing tarantula would enjoy.

Thomas P. Vitale

The scientists and volunteers head back out to the dig site early in the morning. Larry and I stay in town with Melissa and the MPH production team to shoot footage in and around Roswell. My first stop, however, is the International UFO Museum and Research Center.

Although there's alien imagery all over the town of Roswell, the museum is in the heart of the area with the most alien "stuff"—in a three- or four-block area, there are tourist shops, diners and other stores, nearly all with a prominent alien theme. The museum was founded in 1991 by Walter Haut and Glenn Dennis, both of whom still keep an office there. Mr. Haut's daughter, Julie Shuster, now runs the museum, which attracts a couple hundred thousand visitors every year. I have the privilege to speak to Ms. Shuster and to Mr. Haut, who tells me firsthand of his experience in 1947 and what he knows about the incident.

After a couple of hours talking and studying in the museum, we meet up with the production team again. They've just come from a couple of revealing interviews with people who were speaking on the subject of the Roswell incident for the first time. These folks are described to me as real "salt of the earth" people who have lived exemplary lives and would have no reason to make up stories. When I see the footage, I'm amazed by what they have to say.

The rest of the day is spent doing interviews and shooting "B-roll" of Roswell for the special, both the alien-themed part of town as well as the "Anytown, USA" part. At the end of the day, we all do some souvenir shopping at the museum's incredible gift shop; I buy stuff for my wife, for other family members and for some office-mates. OK, I buy myself a few things as well. (How could I pass on the alien-in-a-trapeze that now sits on my desk?)

September 23, 2002

Dig Day Three started out with good news: Dave Hyndman's metal detection search had recorded an "exciting" anomalous return at a location Bill Doleman recognizes as being consistent with metallic debris having washed down the slope of a hill and then being buried on the spot. Could this be the Holy Grail of Roswell everyone was searching for?

The frustration among some of the volunteers—whose numbers had dwindled to seven—built to near-mutinous levels, and another "instant lecture" by Doleman on "the meaning and importance of soils and stratigraphy in archeology" didn't help. The volunteers, having come from far and wide to find pieces of a spaceship, were sore and still chafing at what they perceived as scientific formalities that were restricting them from not only finding, but looking for the object of why they were there. Fortunately, the decision to sink new test pits in the potentially fruitful location found by Hyndman's survey diffused another confrontation, and the first HMUOs of significance would be found that afternoon.

Meanwhile, other project activities proceeded apace. SCI FI's Larry Landsman and Tom Vitale were back in Roswell lining up a bank to receive, secure and store the bags of soil and recovered HMUOs until such time that they could be properly analyzed (which wouldn't be until some months hence). Don Schmitt and Tom Carey were at the dig site to meet a helicopter that was coming in from Albuquerque. Their research had led them to conclude that there were three "crash" sites involved in 1947. Two had been located—the site of the attending archeological dig (known as the "debris field" or "skip site") and the "body site" that was 2.5 miles to the east. The third, the "impact site," is where they believed the remaining portion of the craft and remaining crew ultimately perished. From the air they hoped to determine the impact site's location, which had been based upon sketch witness accounts. MPH Entertainment was also there to film some harrowing moments in the air and on the ground, as well as the project's first casualty.

Dr. Bill Doleman

Dave Hyndman comes out about 11 A.M. to show me the color map of his metal detection (MD) results and the "very exciting anomaly." It looks just like what would happen if metallic debris had been washed down and then buried in one spot. We immediately start setting up new test pits in the MD anomaly, and as soon as the volunteers reach the bottoms of the current "drainage bottom" pits, we put 'em to work there.

The MPH folks were filming interviews in Roswell all day Sunday, but are back with us today. Bob, Lou and I alternate between getting a few mapping shots and coaching the volunteers and setting up new grids in the MD anomaly area. I'm down one volunteer: Jerry, who had to return to work. Lane, bless her heart, has persevered bravely on their pit and made great progress. By now all the volunteers—Debbie and Chuck, Lane and Jerry, Harold and Kate, and Nancy and John— are practically seasoned field veterans needing little supervision. I'm very proud of them.

By afternoon, everybody is busy with something. MPH had arranged for a helicopter to film the project from the air and to take Don and Tom to some of the other crash-related sites. I beg and beg for a ride; I've done the helicopter thing before, and my reason for going was to get stereo photos of the site from the air—you know, like the ViewMaster pictures? At first there's hope, but then there's no room. Oh, well.

At lunch I give the volunteers a 25-cent lecture on the differences between the various aspects of "dirt"—that is, between "deposits" or "sediments"—and the layering in them that we call "stratigraphy" and "soils," which are the horizons that form within the deposits. The characteristics of soils reflect the action of long-acting natural processes, including decay of vegetation into "humic" materials, and leaching of various minerals—especially calcium carbonate and iron—from the surface layers and their precipitation into the lower layers. I explain how these soil-forming processes continue to act over long periods unless erosion or more deposition buries or removes the sediments. Thus, the intensity of the resulting soil horizons offers a rough measure of the sediments' ages.

My lecture is brilliant, if I do say so myself. Too brilliant. My educational effort backfires as soon as the volunteers figure out they've been digging in deposits that are in all likelihood older than 1947, the year that's the focus of our efforts. Discussions and a near-mutiny ensue, though they're happy to start new pits in the metal detection anomaly area. And now they're so good that they have their new pits down to four or five levels by day's end.

Once more we transfer all recovered materials (HMUOs and soil samples) to the security van under the close supervision of FSI's armed personnel. We return to Roswell for another great dinner hosted by SCI FI, and then finally back to the motel. I call Alley, who has returned from California (his team lost the race, owing to a blown engine), and give him directions to the site. I watch an old *Twilight Zone* on SCI FI Channel until I fall asleep.

Thomas J. Carey

This is the day that Don and I go up in the helicopter to try locating the final impact (crash) site of the 1947 UFO. It will be my first trip up in a helicopter, so I am excited and apprehensive at the same time. Don tells me that when helicopters lose power, instead of gliding down like an airplane, they fall like a rock.

I didn't need to hear that.

MORNING

Out to the Foster Ranch archeological site again. The dig is still in progress, but our cast of volunteers has dwindled to seven. Larry, Don and I meet with Bill Doleman to consider another strategy to move earth if the SCI FI Channel will fund it. We decide it is time for Alley Cat, the world-renowned backhoe operator who has scooped tons of earth far and wide. SCI FI agrees that this is a good thing to try. In effect, it is a "Hail Mary" pass to process more soil than we are now doing with our diminished crew. Alley Cat will be brought in tomorrow, the last scheduled day of the dig.

I hear the faint swoosh of helicopter blades in the distance and know immediately that this is it. We have prepared months—no, years—for this. We have known for quite some time that the currently accepted impact site for the 1947 crash is bogus, having been the creation of one Frank Kaufmann for reasons only he knew. In his final days, Kaufmann acknowledged this fact to us, but we already knew. The information we have in hand—descriptions, maps, topography— was given to us by a number of people who claimed to have been to the real impact site at various times over the years. All but two of them are dead now, and the two who are still alive are no longer cooperating. Otherwise, we would have brought them along to guide us.

The helicopter sets down; I thought it would be bigger, but hey, wc got one. MPH wants to shoot Don and me entering the helicopter for the flight. On the first try, I fall flat on my face as I step through the helicopter door, and almost roll out the door on the other side. MPH wants another take. This time, I am able to enter without incident.

The compartment is small and cramped as Don and I take our seats; then I learn that we have to fit Mclissa, the video cameraman and the sound man into the tiny compartment as well. I can't believe this. There's no room to move, and barely enough to breathe. We are all fitted with helmets and intercom phones. It takes about 10 minutes to figure out all of this. Finally, up we go.

We have predetermined a search area based on the testimony of alleged eyewitnesses. In short, we are looking for a flat area with an old, downed windmill on the Jerry Martin ranch, about 30 miles due east from where we took off. The crash site is alleged to be within 100 yards of the downed windmill, with the ground supposedly still disturbed from the 1947 crash. Black rocks are also supposed to dominate the area.

Within a few minutes of being airborne, I start feeling sick to my stomach; the combination of the helicopter's motion and oil fumes seeping into the seating compartment (the engine compartment is right behind my head) is making me queasy. I am surprised at how small things appear from the air, so we direct the pilot to go lower.

We spend about an hour and a half in our search, but do not see any downed windmills (we'll later learn that all downed windmills nearby were removed within the last few years). We decide to head back to camp and try again after lunch.

Just then, the pilot informs us that a warning light has come on. "This has never happened before! I'd better set her down," he says. We set down in the middle of the desert, in the hot sun. The pilot's making sounds like he can't fix whatever's wrong with the engine. We are there about an hour, and I am still feeling sick. The hot sun is making me feel light-headed as well. If we cannot get airborne again, it will be hours before anyone can reach us in the middle of nowhere.

The pilot decides that the warning light must be a false alarm. We are happy, though apprehensive about taking his word for it, but we all clamber back into the helicopter and hope for the best. We are up in the air again (I wish Don hadn't told me about helicopters falling like rocks) and reach camp without incident about 20 minutes later. Still feeling sick, I can barely stand up, and all but collapse after exiting the helicopter. The combination of the dig site's altitude (6,000 feet), the hot sun, dehydration, dust and the helicopter's oil fumes has finally gotten to me.

I am carried to the RV and told to lie down on the bed. Dr. Granek is called to the RV; he thinks I might have a case of heatstroke, so I am packed in cold towels for a few hours. This causes me to miss the afternoon helicopter flight, which renews the effort to locate the Martin Ranch impact site. Schmitt will have to go it alone. I spend the rest of the afternoon flat on my back.

The helicopter returns; Don says he found a site that seems to match what we are looking for, so that's where they conducted the interview. Sounds good. Melissa has lost her cell phone. Good luck finding it (though I think she does, later).

We head back to Roswell. I eat in and spend the night in bed, totally spent.

Donald R. Schmitt

Well, today we're finally airborne, on our long-awaited goal to pin down the final crash site and find that elusive wooden windmill. As it turns out, our pilot once flew helicopter for TV's popular series *Rescue 911*. Before the ride, Melissa wants to shoot some preflight activity on the ground—us exiting from the RV, reexamining our satellite photographs, then heading to the copter which has just completed its checklist.

Our pilot, Tom and I wait for our cue from the director. "Action," she shouts. The RV door won't open. "Film rolling, action." Still no exit. Finally, someone from the outside has to let us out, which gives everyone a good laugh. We next move on to the cramped quarters of our waiting ride. All of Tom's ballroom dance lessons pay off as he performs a headfirst pratfall while attempting to take his seat. I sure hope they save all these outtakes.

Later on, the helicopter touches down to switch the cameraman to the front cockpit glass bubble, but a flashing warning light prevents an immediate return to the air. After repeated efforts by our pilot to ascertain the problem, one of his ground crew suggests we chance flying back to the dig site. Upon returning to our point of departure, we quickly learn of the anxiousness of everyone waiting on the ground.

The exact panicked message shouted from one end of the excavation site to the other was, "The helicopter's down!" It was like yelling fire in a crowded theater.

Debbie Ziegelmeyer

As usual, we meet at Wal-Mart around 6:30 A.M. The day starts out cold, but soon gets blistering hot again. By now we've all had little sleep over the last few days; our muscles are sore, and our necks and arms are sunburned. Plus our dad is in the hospital, so Chuck and I are wondering if we'll have to leave a day early. We work through lunch; there's a lot of ground to cover, and time is running out.

Today Don Schmitt, Tom Carey and the SCI FI crew are in a helicopter for aerial surveillance and overhead shots. By the end of the workday we're so tired that we're all a little punchy. While the helicopter flies overhead, I, Chuck, Nancy and John (who are husband and wife) lie on the ground and form a human frame for our large square hole. I know the helicopter crew must think we've lost our minds; maybe we have, just a little. Next we take turns standing in our knee-deep holes and take pictures of each other as we try to figure out which actors SCI FI would replace us with if this becomes a movie. The group decides that Chuck and I would be played by Donny and Marie Osmond, and Don would be played by Sylvester Stallone.

The day's work is productive, though; Chuck and I find our first HMUO (pronounced ha-moo-o). It's bagged, labeled and handed over to Luey for safekeeping in the armored truck. At the end of the day it's back to Roswell we go, for another late-night dinner and an even later discussion between Bill, Chuck, Bob, Luey and myself. Well, mostly Bill and Chuck.

Chuck Zukowski

Monday morning. We motor to Wal-Mart by 6:30 A.M., where SCI FI is back to join us, and the camera crew is ready to roll. Today is destined for excitement.

When we arrive at the debris site, there are no instructions; we're all well versed in procedures and start our morning routines. Tarps are pulled, pictures taken, sifting tripods raised, knee pads on, tools in hand, and we're back to work. Cameras in the background, interviews running amok, and then the helicopter flies in. Yep, this was going to be an interesting day.

Tom Carey and Don Schmitt start patrolling the area like field marshals, making sure every one of their whims are being fulfilled. My sister and I are doing well with our dig, but we're becoming frustrated with any type of production delays. We opt to skip lunch and stay with our dig; we don't want to miss one bucket of dirt searching for that Holy Grail of Corona.

While everyone eats 100 yards away, Debbie a   own thoughts as to what has been transpiring at th   dict what Dr. Bill was doing, or what Tom war   personally running out of time. Being vol   paid; in fact, we're losing money every d   Debbie runs her own business with her husba   design microchips. Therefore, no work, no pay. S   portant, and a must to do.

With our previous plans to leave after the dig tonight   we don't want to miss one single moment of daylight. Vicious   ing, we're able to start on another site. This one was derived from   cussions between the archeologists and us, and was in an area smack dab in the middle of where anomalies were located underground. Wiping our brows and pulling up our already pulled-up shirtsleeves, we dive into work. While others munch on food, we munch on dirt. Having already consumed my yearly requirement of minerals over the last couple of days, I begin working on next year's.

Lunch breaks and the other volunteers migrate back to their digs, once again working their sites as if they were gold miners from days gone by. Off in the distance, a whining sound tells us the helicopter is making ready for flight. Don, Tom, SCI FI cameramen and a host of others are filing into the bird. Aerial investigation was the priority, shooting pictures a close second.

A gust of wind, and the chopper's gone, heading east for new locations—possibly the crash site, possibly the site where the bodies were flown out. Only Don and Tom know, while we stay behind to guess.

An hour or so passes when one of the crew comes running down to my site, yelling my name. I meet him partway, unsure of what was going on.

"Chuck, do you have your GPS with you?" he asks.

"Sure," I say. "I don't leave home without it."

"Could you find a downed helicopter if you had the coordinates?"

My heart slows. "Sure," I say. I grab my GPS from my investigator kit and run toward the helicopter's fuel truck. A burly man stands there, radio in hand, in direct communication with the helicopter pilot. Everything's all right; there are no injuries. A warning light had alarmed the pilot, and he made a quick decision to land. It's unknown if they can take off again.

Once again I'm asked if I can locate the helicopter if I'm given coordinates. "Yes," I say. "I found the debris site over a year ago with this thing. I can find anything with valid coordinates." The mechanic hands me the coordinates, which I immediately enter into my GPS, stopping only to recheck the numbers he has given me.

"The pilot says he's about 19 miles away," says the mechanic.

"Actually he's 22.9 miles due east," I say. I could be there if need be. While the mechanic and pilot decide how to proceed, I ready our rented SUV in case we need to roll.

Fifteen minutes later, they decide to try a liftoff. If successful and without warning lights, the chopper would head right back to our base. Up in the air it goes, while we wait. A "thumbs-up" from the mechanic. Chance favors SCI FI and the investigators on board; this is indeed their lucky day. A simple wave of my hand and tip of my dusty cap, I go back to my site.

Later in the afternoon, with the teams of volunteers all digging within a 30-foot radius, we're running like a fine-tuned Mustang and finding artifacts. We're making progress, while the helicopter again makes for the air, this time for shots of our work around the site. We dig through the ground, sift through rocks and pebbles, and don't do anything out of the ordinary from what we've done all weekend. It's nice to know we're being filmed; jokes are made, sly remarks answered, and smiles come to the volunteers.

The smiles soon turn to frustration, however, as the helicopter makes another pass, and another, then another. The right shot, the perfect angle . . . we've become props. We work on, but decide a little mutiny is called for. The next time the chopper flies over, I, Debbie and a couple of other volunteers have some fun—we lie down around one dig site and create a human square.

We don't know if Don Schmitt had instructed the pilot or if there was a downdraft, but the next flyover is a hell of a lot lower. Kicking up dust, they have the last laugh. On purpose or by accident, it's still funny and we get a kick out it. The SCI FI people are very professional, but still human. We've poked fun at each other all through the dig—a jab here, a jab there, a finger in your nose while they're shooting your picture (I hope they don't use that one) . . . I'm sure we've made some good friends.

After a long day, we head back to Roswell. Debbie and I are asked to secure reservations at a local restaurant; after a quick cleanup in our motel rooms, we wait for the crew to arrive. Our plans are to eat, shower and hit the road for home, yet something still bothers us about our dig, and Tuesday would gain a backhoe at the site. Should we stay another day, reschedule flights and take another day from work? Time would tell, and that time is right after dinner.

After eating, I make the decision to once again stick my nose where it doesn't belong. I walk over to Dr. Bill and suggest an alteration in the way my sister and I would dig at the site. If he'd allow us the opportunity to run our own dig, we would stay one more day. Other volunteers were leaving, and only the backhoe would be working on Tuesday.

After much discussion, it's decided that if I want to strip-excavate, I would be allowed to. Strip excavation was taught to me while working at another archeology site in Arizona. I wanted to work one-meter-by-five-meter strips, digging down about 10 centimeters, or just far enough to find something hidden 55 years ago. By the grace of Bill, Debbie and I would control our own dig tomorrow, which would highlight our excellent adventure.

Thomas P. Vitale

As the dig team and production crew head back out to the site, Larry and I go around town looking for a bank that will let us use its vault to lock up the bags holding the HMUOs and soil samples. We meet with the bank manager at Wells Fargo and make arrangements.

I take a midday flight out of Roswell back to Albuquerque, on the same 19-seat Mesa Airlines prop plane in which I had arrived. As we wait to take off, I look out my window at just the right moment; a stealth bomber is flying past the plane! I had seen stealth bombers on television and in movies, but never in person. I immediately move to the other side of the plane and watch as this incredible jet quickly recedes from view into the distance. Watching the stealth fly, a plane that looks like no other on Earth, I can't help but wonder if it was developed via human ingenuity or back-engineered from alien technology. (Hey, four days in Roswell will do that to you!)

I wish I had the time to stay until the end of the dig; unfortunately, I have to get back to the SCI FI office, where other projects are pending. Still, I come away from my experience with much greater knowledge of the events at Roswell. I now understand why this incident has experienced such amazing staying power in the public imagination. It's an incredibly compelling story, with more and more bits of the truth being pieced together all the time.

My personal feeling is that if it was an alien craft that crashed there, it's time the public knew the truth. If the explanation is more prosaic and Earth-based, then people still have a right to know. I don't believe the government's explanations of the incident jibe with all of the evidence. My strong hope is that our special, and the work SCI FI will continue to do in the area of UFOs, will help this topic remain part of the nation's discourse. Perhaps over time, more people will come forward with whatever truthful information they have, whether or not that information supports a belief in UFOs.

As Fox Mulder would say, the truth really is out there, whatever that truth is, and humanity has a right to know.

September 24, 2002

Fourth (and final) Dig Day. Due to severe time constraints as well as the departure of most of the volunteers (only four remain at this point—two diggers and two sifters), the day's archeological activities can best be described as "Wing-It Day" or "Quick-and-Dirty Day." Since not enough earth had been moved during the three previous days' digging, it was decided to bring in a backhoe to do some trenching in high-potential areas, including the gouge area and the area where Dave Hyndman found his "exciting anomaly." After a meeting of the minds with Bill Doleman over dinner last night, the remaining volunteers were granted permission to "do their own thing" and conduct shallow strip-digging, or trenching, over a wider area.

Promising results would justify the last day's decisions. "Alley Cat," the

The use of backhoe trenches to search for furrow evidence and to investigate anomalies revealed by the electromagnetic conductivity and metal detection surveys was implemented to increase the chances of discovery, and to allow investigation of depths greater than those possible in 50 x 500 cm test pits.

backhoe operator, was able to dig ten long and deep trenches in the predetermined spots; one across the line of the alleged gouge gave indications that such a feature was indeed present where it was said to have been. The strip-trenching activities by the remaining volunteers produced additional HMUOs, which were bagged for shipment to and safekeeping in Roswell. The contour mapping of the site was finally completed, enabling Bill Doleman to go off on his own and investigate a surface anomaly he had noticed a few days earlier. Bill dubbed the anomaly an "alternate furrow" because it paralleled the track of the alleged gouge and resembled a gouge itself.

While MPH Entertainment filmed the final round of interviews for SCI FI's expected documentary, 80-year-old Frank Joyce, a key eyewitness to the 1947 events, showed up unexpectedly and provided an on-the-spot interview; a real coup, as was the provost marshal's daughter a few days ago. After a final location interview that re-created the finding of the site to be excavated, it was time to head out to Roswell for the last time. Not before, however, some final, harrowing moments on the trip back.

Dr. Bill Doleman

Our last day. I discussed the excavation methods with one volunteer at dinner last night and explained that we weren't digging a known site, but testing one—a fact that makes a big difference in methods, and was one reason why we were using the geophysical prospection, plus our knowledge of natural processes, to guide us in where to dig. I also explained that since burial by animals was one possible way in which any debris missed by those who reportedly "cleaned" the site in 1947 might have been preserved to the present day, it was important to dig deep.

Nonetheless, we agreed that we had a pretty good handle on the site's stratigraphy. If there was erosion-buried debris under the surface in the metal-detection anomaly area, it would probably be pretty shallow. It was time to try an alternative methodology: shallow stripping of wide areas.

Alley's supposed to meet us at the turnoff from the main road, but he's not there. As soon as we get to the site, Lou and Bob start setting up a long, one-meter-wide stripping trench across the MD anomaly. While the volunteers wrap up the deep pits they excavated yesterday in the MD anomaly, I go back to the main road and find Alley; he had gotten to the turnoff just after us, flagged down a local and asked where the archeologists were. Her response? "Oh, you mean the ones digging up the UFOs?" He said yes, and she pointed him in the right direction. So much for project secrecy!

I lead Alley into the project area; he and his 25-foot trailer with backhoe move very slowly across the rutted two-track road. I park at the top of a hill where he can see me, then wander off in search of a curious feature that my crew and I had noticed every day as we drove in. What did I find? Our nondisclosure agreement prevents me from saying until the special airs on SCI FI November 22.

Nancy and John had decided to return to Colorado, but the volunteers who remain spend the rest of the day stripping the long trench. MPH is conducting interviews of them, and of some well-known Roswell incident insiders who had come out to the site. Frank Joyce is here, along with Don and Tom; even one of the volunteers was the child of a major player in the July 1947 events. I get all their autographs!

We get some more mapping done, and I lay out a number of

To prevent any "salting" of the site, excavation units are photographed on a regular basis, including the natural surface of each test unit prior to beginning excavation, the grid's surface at the completion of each excavation level, and the grid's surface at the end of each day prior to covering.

trenches for Alley to excavate, including six across the furrow/gouge centerline (I have Don point it out to me again, to be absolutely sure). I also lay one out across the southern electromagnetic conductivity anomaly (which Dave Hyndman said no amount of data processing would get rid of; it was "real"), then another across the MD anomaly, parallel to the stripping trench in which the volunteers are digging.

I go off to inspect some of the site's geology, and to take a look at what Alley is exposing in the trenches he was digging. At the same time, I catch up on my photography of various test excavation areas, geological features and what we thought were test pits that were placed in the site by some archeologists in 1989. These pits tell me two things. First, we were probably at the same place folks had previously, and briefly, looked—i.e., their eight pits were shallow, and not mapped or reported. Second, the pits looked to have been no more than 10 to 20 centimeters deep, to have been left open and to have been filled in partially by natural erosion. They were also partially overgrown.

These latter observations offer crucial evidence. The amount of infilling and vegetation growth in the 11 years that had passed would give baseline information on the rate at which natural processes might have obliterated the 1947 impact evidence through burial and vegetation growth. I photograph a good sample of them, as well as some geological features I was interested in. It was getting close to time for my interview, so I decide to check the trenches Alley had dug across the furrow centerline. As to what I see there . . . well, let's just say the words "Holy cow" come out of my mouth. I take more pictures, then go off to my interview, which lasts until just after sunset. I don't say anything about what I've found quite then. When I'm interviewed later I'll explain what it was and why it surprised me, but for now that pesky nondisclosure agreement . . .

The volunteers wrap up under Lou and Bob's supervision, and the last of the HMUOs and soil samples are transferred to the security van. The van will return to Roswell, where everything will be locked up until decisions are made about how and where to analyze the materials (not up to me—I'm just an expert field technician whose job it is to look for evidence using appropriate methods). Bob and Lou begin gathering up equipment to put in our vehicles for our return to Albuquerque.

At last we pull out. It's now almost totally dark. I'm in a hurry to get back to Albuquerque and to my lovely wife and two dogs. I arrange for Jeff, one of the MPH crew, to take several people back to town so I can head straight home. I say my goodbyes and start to leave, but then decide I'd better make sure everybody got out okay. I'm glad I waited; everybody else is doing fine, but Jeff has a rock stuck in the transmission linkage of his ancient sedan and can't get it out of first gear, even to park it. I follow him and his passengers to the main road.

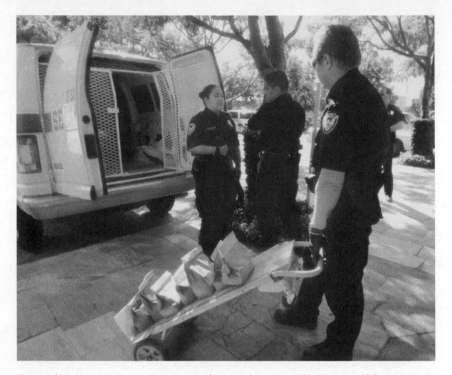

Extraordinary security measures were implemented to ensure the integrity of the project. At the conclusion of the primary phase of the dig on September 25, 2002, fifty-three bags of soil samples and twenty-four envelopes of HMUOs (Historic Materials of Uncertain Origin) recovered in the course of the excavations were transferred to lock boxes at the Wells Fargo Bank in Roswell.

Jeff gets the rock out, but the car's still limping along; the fuel line has also taken a hit from a rock and is leaking slowly. The passengers climb into my van and we all head to Roswell; Tom and I have a much-needed cold one while everybody else "crashes." I'll just sleep in and shop for souvenirs at the IUFOMRC and other nearby stores before heading back to Albuquerque. It has been nine long days in the field. I'm excited, exhausted, and ready to come back out and investigate certain things further.

Thomas J. Carey
9 A.M.
Still feeling a little woozy from yesterday, but we have an appointment to meet Frank Joyce at a ghost town named Lon. It's located

on Highway 47, between Corona and Highway 285, about an hour or so out of Roswell. Since all that is allegedly left of Lon are the building foundations, we have not been able to locate the town ourselves, as we have driven by it on our daily trips to and from the dig site.

The reason that we want Frank to show us Lon is because in 1947 the military told his boss, KGFL radio station owner Walt Whitmore Sr., to take him to an abandoned building in the town. It was there he met with Mack Brazel, who told Joyce never to talk about the Roswell Incident. We want to photograph Joyce at the exact spot where it happened.

I know we are in trouble when Don says we will meet Joyce there (despite our not actually knowing where "there" is), rather than meeting him first and driving out together. Well, my concerns turn out to be well-founded, and we never hook up with Joyce. We finally give up and drive on to the dig site. I almost roll the vehicle at a muddy spot on the narrow almost-road; for a few seconds, I have absolutely no control of the skidding vehicle.

Once at the site, MPH interviews Don as they interviewed me on the first day of the dig. While he's being interviewed, who should drive up but Frank Joyce! "Where were you guys?" he wants to know. I think I must be dreaming; we drove all over the place searching for him. And how did he know how and where to find us? We still don't know. I point Don out to Frank, and we go down to talk to him.

Alley Cat and his backhoe are in full operation. How did he get it out here to such a remote place over such rough terrain? You can't beat experience.

While Don and Frank are occupied, I ask Bill Doleman for some digging tools and take off on foot toward the east, away from the main area. My plan is to do some quick and dirty digging in the outlying areas, away from where the wreckage was thought to be back in 1947. My thought is that the wind might have blown some of the lighter pieces of wreckage some distance away from the main concentration. Maybe I will get lucky.

I don't.

After an hour of digging without success, I head back to the main area. On the way, I run into Frank Joyce, who is on his way out. I thank him for coming and apologize for our earlier missed connection.

It is late in the afternoon, and the sun is starting to go down. The backhoe has stopped digging. I'm ready to leave, but I now learn that MPH wants Bill Doleman and me to re-create the first day we took Doleman to this site, to get our bearings on where the gouge might have been. How long is this going to take, especially since Doleman seems otherwise occupied with being interviewed by MPH?

Now Don tells me that he has to get back to Roswell on another matter; there goes my ride back to town. I'm told that a call has been placed to one of the MPH gofers to drive out here and take us back to town when we are done. He apparently was seen an hour ago as he neared the dig site, but no one has seen hide or hair of him since. I also hear that he's almost out of gas. This is not looking good.

Don is now gone, and I have been marking time for an hour and a half to do this re-creation scene with Doleman. There is hardly any sunlight left, but we finally do the shoot.

We head back to camp to pack up and leave for the last time. I learn that the two MPH cars are not going back to Roswell, but are going straight to Albuquerque to do another interview. Ditto the photographer, Babak. That leaves the archeology van, though Bill Doleman says he is also driving straight back to Albuquerque from the dig site and not passing "go."

Wait a second! Someone spots a dust trail on the horizon. It's the MPH gofer. He's finally found the site, after wandering in the desert for an hour or so. But wait! What's this he's driving? Nothing less than a 1978 Oldsmobile Cutlass that's badly in need of a paint job! How in God's name are we supposed to get out of here in that? What dope dreamed this up? And he is out of gas! This surely must be a dream. Please, let me wake up.

Gasoline is siphoned from Alley Cat's rig into the Oldsmobile using a rubber hose—just like in the old days.

Another surprise: three more people will be joining us in the Oldsmobile for the ride back to Roswell. So, on top of everything else, we will make like sardines for the entire trip. Sure enough, the car bottoms out at the first rough spot, sounding like the *Titanic* as it scraped along the iceberg. "The gearshift won't move," exclaims Gofer. Someone in the back seat shouts, "I smell gas." We all pile out of the car.

Fortunately, we are first in the convoy of vehicles, otherwise we'd still be out there. To my surprise, though it is obvious we are in trouble, the two MPH vehicles and Babak the photographer go around us and wave as they speed off into the darkness, leaving us here.

After a few minutes, the UNM van, with Bill Doleman driving, arrives. Gofer and one other from our contingent have been underneath the Olds trying to loosen up the steering, but to no avail. Doleman asks Gofer if he thinks the car will make it to Roswell. "I think it will if we drive at 20 miles per hour," he replies. At that speed, we will arrive in Roswell just in time for Christmas.

Gofer goes back to the Olds, and Dolemen grabs the other fixer-upper, Bob. "Quick, tell me, Bob. Will it make it?" he asks.

"Not a chance," is Bob's prompt reply. I head back to the Olds.

After a few minutes, I notice that it has gotten really quiet. Too quiet. No voices. I get out of the Olds and head back to the UNM van. Doleman isn't behind the wheel, and no one's inside. Strange. What's going on? I head in back of the van, and there stands Bill Doleman, his sidekick, Bob and Gofer too, all with their arms folded across their chests, staring up at the stars and not saying a word. Surreal, indeed. Here we are, stuck out in the desert in the middle of the night, and they're looking at stars (looking for help?). I really must be dreaming now; this cannot be happening.

I look up myself, and the starry sky seems like the kind you see in planetariums—certainly not the sky that one sees back home near Philadelphia. I can actually see the Milky Way (for the first time!), and it looks just like it did at the Fels Planetarium. In spite of our plight, I am impressed.

10:30 P.M.

It's decided that we should all travel in the UNM van. Bill Doleman graciously agrees to take us back to Roswell; he will stay overnight there and drive on to Albuquerque in the morning. Gofer decides to take a chance on driving the '78 Olds Cutlass back to Roswell. We decide that it would be best if we follow Gofer in case his car breaks down—something we expect to happen at any moment.

On the drive back to Roswell, one of the passengers yells out, "There goes my husband!" He is heading in the opposite direction up the highway. Turns out that he had been waiting for his wife, a fellow passenger, since 6 P.M. at the location where she was supposed to be dropped off. When she didn't show after a long wait, he decided to head on up the road—all the way to the dig site, if necessary.

11 P.M.

We get back to Roswell without further incident. Our passenger's husband also arrives to pick her up; he isn't looking too happy, and who can blame him? I am exhausted and thirsty. So is Bill Doleman—at least the thirsty part. After a few tries, we finally find a place that will let us in to have a beer, but only on the promise of leaving a good tip (which we do). Afterward, I thank Bill for his skill and professionalism throughout the time of the dig, and especially for helping us out when we were stranded this evening. A good man.

I hit the sack about midnight after winding down from a harrowing night.

Donald R. Schmitt

Our last day at the dig site starts with us driving to meet Frank Joyce, a friend who was a newsman for Roswell radio station KGFL back in 1947. Frank was driving down from Albuquerque to show us where the late town of Lon once existed. (Lon had played a central part in the aftermath of Frank's personal experience.) We become concerned—after waiting at our meeting location for almost an hour, Frank has yet to arrive. We proceed to the Corona debris site, which is devoid of marked roads and landmarks. So imagine our surprise (and our relief) when who should arrive an hour later but good old Frank Joyce, unescorted! Though it had been more than 50 years since he had ventured to the now-famous site, the area had left quite an impression on him.

The site is a flurry of ongoing activity, which makes my interview a bit distracting; noise is a constant factor, but I think it goes well. The crew talks with Frank before heading up to Albuquerque to interview some CIA agent, who I'm sure will tell them that all of our witnesses are liars and the government has never covered up anything. Debunkers are such cowards; it takes absolutely no guts or effort to disbelieve everything.

I race back to Roswell later that afternoon, and I do mean race—I still have urgent matters to attend to, which includes finalizing all of our plans to secure a scan of the General Roger Ramey memo negative. We're going for the highest DPI available and hope to computer-enhance the image during the process.

It's getting late—after 10 P.M.—and still no sign of Tom. It's not like him to miss dinner. I can just picture him still up at the site, after dark, feeling like he's the last person on Earth. Hope they're getting this on camera.

Kate Leinster

Most of the volunteers have had to leave, leaving only a few of us. After a "friendly debate" last night between Chuck and Bill, our digging process is altered in the hopes of maximizing our HMUO finds. Thankfully for Chuck, it does, so go, Chuck! While he and Debbie decide their exact digging strategy, I use the time to do some university revision; I have an exam the day after I get back.

Leaving the site for the last time makes me feel somewhat sad. I know that this is not somewhere I will probably visit again . . . unless, of course, the HMUOs are found to be extraterrestrial and we need to dig further! All the hard work was definitely worthwhile—some interesting things have been found, and I am anxious to hear the results of the testing. I've certainly had an adventure and met some lovely, weird and interesting people along the way. I head back to Arizona feeling grateful that I have been part of something special and unique.

Debbie Ziegelmeyer

Today is my daughter's 27th birthday. I am in New Mexico and she is home in Missouri. Dad is out of the hospital, so Chuck and I have decided to change our travel plans and stay. We arrive at the site around 8 A.M. and—with Bill, Bob and Luey's help—stake out a new work site.

Bill Doleman and Chuck had looked at the ridge, some charts and the water flow, so today we're using a different excavation approach in our search for HMUOs. This being our last day, Chuck and I work through lunch again to maximize our search efforts. Today's excavation volunteers are Chuck and me. End of list. Around noon, Kate and another lady help me with the sifting; poor Chuck is still on his own with a lot of ground to cover, armed only with a shovel.

Don Schmitt shows up on the scene, grabs a shovel and gives Chuck a hand for a while. Though he was dressed nicely for an on-camera interview on the site and some afternoon meetings, he still rolls up his sleeves and starts digging. We really appreciate the respect he shows for our efforts by doing this. By the end of the work day, we have again bagged some very interesting HMUOs—one in particular—and are pleased with our efforts.

We find it hard to leave with still so much ground to cover, and even harder to leave our new friends, but we say our goodbyes and depart. We don't have a hotel to go back to for a shower, or the time, so we drive away from the site tired, dirty and hungry. We have to make our way back to Colorado Springs, which is about a seven-hour drive. Chuck has to go to work first thing tomorrow morning, and I have a 9 A.M. flight back to St. Louis.

Chuck Zukowski

Debbie and I meet Dr. Bill at Wal-Mart at 6:30 A.M.; I think he's surprised to see us this time. We make our usual stop at the gas station for coffee, and then head off to the site.

Dr. Bill has some difficulties contacting the driver of the backhoe, which should have been near the highway when we arrived. I'm sure the driver's lost; hell, it took GPS coordinates to find this place! Debbie hands Bill her satellite cellular, and he's soon able to contact the driver and vector him in. All is well.

Immediately afterward, Bill starts working on our new site. Accounting for some underground anomalies, the curvature of the sloped hill, the contour of the ground and the estimated erosion from 55 years of rainfall, Bill flags out a section one meter wide by about 25 meters long. One of his staff promptly measures out the section in one-meter increments. After laying out guide twine with a flag marking every fifth meter, Debbie and I are ready to start.

I move the sifting tripod and our digging equipment over, then begin moving the earth with a flat-head shovel. Like a finely tuned clock I'm filling buckets of dirt while Debbie sifts it through the screen. Off in the distance, the backhoe arrives, and a couple of other volunteers show up about the time SCI FI arrives. Debbie and I stay with our dig through lunch, and I soon use other volunteers to control two other sifting tripods. With three sifting racks moving while I feed dirt to them, we become a well-oiled four-cylinder engine; you can hear the motor running.

Once lunch is out of the way, SCI FI sets the cameras up in front of our new location and uses us as the backdrop. By that time Debbie and I have uncovered, bagged, logged and secured some very interesting HMUOs. We're pretty happy with our progress. Don Schmitt is in the background talking to the cameras, people we've never met before are walking over to us to get into the shot . . . we're in complete control of our situation.

SCI FI's excellent photographer, Babak, comes to our exotic dig, says, "Hey, this is new," and starts taking pictures. So now we have multiple one-meter by one-meter dig sites overshadowed by a few one-meter by five-meter strip sites cutting right through them. Cool. Even Don Schmitt comes over and helps with some digging. Well, just a little, but it's nice to see his approval of the new, unusual dig. Tom Carey swings by and is interested in what we're accomplishing; after a brief explanation, he too is satisfied and on his way. There's still a lot to do; time is of the essence, but we're kicking butt.

By the time the last ditch is dug, it's just Debbie and me again. After losing our other two volunteers to other duties, we finish the dig and call it quits. We dust off our clothes, pick up our personals and return to the SUV. We run into Don on the way there, brief him on our successful day and head toward Dr. Bill.

Now, Bill Doleman and I didn't see eye-to-eye at times. Using science as a tool, he was very good at what he was doing. Using research as my tool, I added some flavor to his thinking. The bottom line? We leave good friends. I highly respect this guy and his team, and honestly believe SCI FI could do no better. I would love to work with him again, and feel confident in what he does.

Debbie and I bid Dr. Bill farewell and head back toward Colorado Springs. With a long drive ahead, we're dusty, tired and sore, but the past few days have definitely been the most excellent adventure we've had. Yet.

September 25

Thomas J. Carey
Last day. I have an 11 A.M. flight on Mesa Air out of Roswell. I will connect in Albuquerque for a flight to Phoenix, and then on to

Philadelphia. Don has driven to Albuquerque and will fly home from there. I finally see my rental vehicle again. Todd told me he'd have it washed before he gave it back to me. Well, he didn't, and he left me an empty gas tank as well. What a guy.

On my way to the airport, I stop off at the UFO Museum to say my goodbyes. At Roswell Airport, I check all of my bags through to Philadelphia except my one carry-on—a briefcase with my notes and assorted reading material.

After arriving at Albuquerque International Airport, I meet up with Don. We go over our just-completed project, as well as what we still have to accomplish in the near term. All in all, we're happy with what has just taken place. All of the witnesses showed up to be interviewed—several for the first time anywhere—and the dig came off as planned. Unidentified items from the dig have been bagged, tagged and sent away for safekeeping and later analysis. The results of the backhoe are still yet to be determined, as Bill Doleman will be returning to the dig site to investigate the trenches that were dug. Also, the report from the electronic scanning device that was looking for the gouge and any other anomalies is weeks away from completion.

10:30 P.M.

My flights are mercifully uneventful, and I touch down at Philadelphia International. Ahh. It is over, but not quite. I meet my wife at baggage pickup. We will be out of here as soon as my bags arrive. 'Round and 'round they go—everyone's bags but mine! Mine are the only bags that do not arrive with the flight. Steaming, I head into the little office near baggage pickup. The attendant says, "We've had some trouble with arriving bags with this flight for some time for some reason." That's nice, but what do I do now? She says that my bags will be delivered to my home the next day. What can I do except go home?

September 26

Thomas J. Carey

My bags do, in fact, arrive home today. Unfortunately, I discover that my camera, plus all the film I shot on the trip—240 pictures total—are missing. Everything else is here.

A follow-up with America West Airlines yields an incredible result: the airline declines accepting responsibility for the apparent theft of my photographic equipment by one of their employees. It's disconcerting to think that America West employees can "treat" themselves to whatever customer belongings they may desire with no consequences whatsoever. Customer "service" has indeed come full circle.

4

THE POLL, PART TWO

In early September, marketing intelligence firm RoperASW conducted its second survey for the SCI FI Channel. While the first opinion poll asked individuals (randomly selected via RoperASW's OmniTel service) their beliefs regarding UFOs and extraterrestrial life, this second study's questions were oriented specifically toward the Roswell Incident and the government's official explanations behind the case.

Roswell: Weather Balloon or Space Craft?

Americans' Beliefs about the Roswell Incident of 1947

Prepared for the SCI FI Channel

October 2002

Roper Number: C205-008232

Table of Contents

I. Methodology

This study was conducted by RoperASW via OmniTel, a weekly national telephone omnibus service. The sample consists of 1,033 male and female adults (in approximately equal number), all 18 years of age and over.

The telephone interviews were conducted from September 6th through September 8th, 2002, using a Random Digit Dialing (RDD) probability sample of all telephone households in the continental United States.

Interviews were weighted by five demographic factors: age, sex, education, race and geographic region. Weights were applied to ensure accurate representation of the adult population in each of these areas.

The margin of error for the total sample is +/- 3% points.

II. Detailed Findings

Familiarity with the Roswell Incident Runs High

Six in ten Americans (59%) have read, heard, or seen something about the Roswell incident of 1947, in which a flying craft crashed in the desert of New Mexico.

Males are more likely than females to be familiar with the incident. In addition, adults between the ages of 25 and 64 are more inclined to say they have heard of Roswell than are their younger and older counterparts.

Familiarity runs highest among residents of the West, followed by the North Central region.

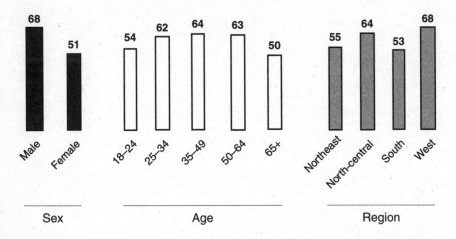

% Who Have Read, Heard, or Seen Anything about the Roswell Incident

Many Americans Skeptical of Government Explanation of Roswell Incident; Most Believe Evidence of Extraterrestrial Space Craft Should Be Shared

Among those familiar with the Roswell incident, four in ten (41%) believe a space craft crashed in the desert in 1947, and another one-third (32%) are not sure if it was a weather balloon, as the government claims, or something else. Only three in ten Americans (28%) who know of the Roswell incident believe the government story of a weather balloon crash at that site.

Males are significantly more likely than females to believe it was a weather balloon that crashed in the desert at that time.

% Who Believe That the Roswell Incident Involved a . . .

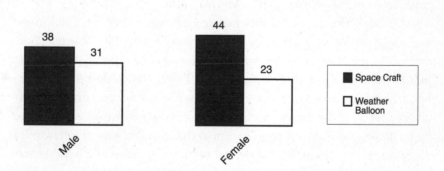

If the government has evidence that it was an extraterrestrial space craft that crashed outside of Roswell, then most Americans (72%) familiar with the incident believe the government should share this information with the public. Agreement with this sentiment is consistent across all demographic groups.

5

THE FINAL REPORT

Those involved with the Foster Ranch excavation didn't expect to find answers to all of the questions that surrounded a mystery more than fifty-five years old. Their focus was on answering just one question: Did something crash down outside Roswell, New Mexico, in July 1947?

Although the dig ended as scheduled on September 24, 2002, there was still plenty of work ahead for OCA project director Bill Doleman. Several factors had motivated the contract archeologist to officially request that SCI FI fund the OCA crew's return to the Foster Ranch skip site in October, where they would conduct soil-stratigraphy studies and finish collecting important site-mapping data. The request was accompanied days later by a follow-up e-mail to Larry Landsman that detailed several noteworthy observations, including what appeared to be the remains of a weather balloon, plus significant differences in soil color within a backhoe trench that Elgio "Alley Cat" Aragon had dug at the impact end of the furrow alignment. These findings offered more than sufficient reason for SCI FI to not only approve Doleman's request, but to send production company MPH Entertainment back to New Mexico and shoot more footage for the network's Roswell special.

....

The University of New Mexico
Office of Contract Archeology
1717 Lomas Blvd. NE
Albuquerque, NM 87131

September 27, 2002

Larry Landsman
Director Special Projects
SCI FI Channel
1230 Avenue of the Americas
New York, NY, 10020

Ref: Contract extension, Foster Ranch Project technical and logistical
support (per UNM proposal 185–769A)

Dear Larry,

With this letter, OCA proposes to extend the Foster Ranch project for
an additional 1.5 days at the regular time rate of $1266 per day as spec-
ified in Sections 1 and 3 of the original contract (total cost $1899.00).
The purpose of the extension is to provide time for myself and two assistant archeologists to conduct soil-stratigraphy studies at the site and
to complete mapping activities. Both activities are necessary to producing a high-quality report that meets professional standards and ad-
dresses the questions targeted by the Foster Ranch project as fully as
possible.

There are two primary reasons for the requested project extension. The first results from the well-advised decision that was made in
the field to incorporate backhoe trenching into the project. The backhoe made it possible to look quickly for the "furrow," but also offered
an opportunity to investigate the conductivity and metal detection
anomalies found by Dave Hyndman of Sunbelt Geophysics and determine if they are natural or "other" in origin. Briefly, the "soil-
stratigraphy" study will help identify the site's natural stratigraphy,
which can then be compared with that found in the volunteer-

excavated test pits, as well as in both the furrow-seeking and anomaly backhoe trenches. This effort will require some more detailed soil-stratigraphic work than I originally planned.

Second, additional time is required because we lost a lot of time whenever the MPH film crew appeared at the site. Time was lost to answering questions and sometimes in waiting for them (e.g., Saturday, September 20, when the MPH crew was an hour late, followed by slow driving time to the site that accommodated the RV's slow speed). As a result, our site mapping activity was slowed. The site map will be an important component of our report and will aid in understanding the results of both the excavation and Mr. Hyndman's remote-sensing studies. Some soil studies during the project were originally planned, but were not possible owing to various interruptions. Other lost time resulted from bad weather on Wednesday, September 18, and extended consultations with Mr. Hyndman in Corona the following day. It was originally expected that those consultations would take place in the field while the rest of the crew worked, but Mr. Hyndman wouldn't risk going into the muddy project area and we conferred in Corona instead.

Please also be apprised that OCA receives no funding from the university and is entirely funded by contracts. Thus, if SCI FI chooses not to fund the extension or the "additional administrative and logistical support" supplemental budget submitted on September 13 (UNM Proposal 185–769B), OCA will incur a loss on the project despite the hard and often uncompensated work (e.g., driving times of 3–4 hours per day) put in on this project by me and my staff. I hope that SCI FI will choose to honor OCA's commitment to the project by funding the requested extension and the supplemental budget.

If you find the proposed project extension acceptable, please confirm your acceptance in a letter to me. And, of course, if you have any questions, please don't hesitate to call or e-mail me.

Sincerely,

William H. Doleman, Ph.D.
Principal Investigator

....

From: Landsman, Larry
Sent: Monday, September 30, 2002 8:54 AM
To: Schmitt, Don; Carey, Tom
Subject: FW: One More Thought

FYI

----------Original Message----------

From: Doleman, Bill
Sent: Friday, September 27, 2002 7:40 PM
To: Landsman, Larry
Cc: Myself
Subject: One More Thought

Larry:

I almost forgot to mention two things of interest.
First is the "other" site I found and described to you
on the phone Wed. I'd like to record it briefly, given
its similarity in nature and orientation, and proximity
to (ca. 0.6 miles) the current site. Said recording
would be covered under the extension I requested today,
as would the anomaly described below.

Second is the apparent stratigraphic anomaly found in
one of the backhoe trenches (don't recall if I
mentioned it or not). The trowel points to it in the
attached photo (note the line running from upper right
to lower left, as well as the differences in soil
colors between the areas to left [lighter] and right
[darker] of the anomaly—sure looks like a possible
"gouge" to me). When I saw it, I photographed it and the
other 2 trenches in that area immediately, but did not
have a chance to look further as I was awaiting my
interview call. This anomaly, found in the trench
located at the initial impact point as determined by
Don S., is right where we'd expect it to be and is
actually pretty exciting. But it clearly needs further
investigation (the soil stratigraphy study) to
determine its nature, origin and implications for our
research goals. This study would also be part of the
requested extension. If you'd like the need for this
particular investigation outlined in a letter, let me

know ASAP, as I will be out of town on a geological field conference from Oct. 2-5.

Have a great weekend!

Cheers,
Bill
=========
William Doleman, Ph.D.
Principal Investigator
University of New Mexico
Office of Contract Archeology
1717 Lomas NE
Albuquerque, NM 87131
=========

....

From: Carey, Tom
Sent: Wednesday, October 02, 2002 2:37 PM
To: Landsman, Larry; Schmitt, Don
Subject: Re: FW: One more thought

Larry:

This is indeed truly exciting in as much as the Air Force is on record as having eliminated all possible causes for the Roswell event except a Mogul balloon, and Charles B. Moore himself has also stipulated that a Mogul balloon could not have caused the gouge as described. We, of course, have eliminated all possible causes (that make sense) except a UFO, and our hope for the dig just completed was to find:

1. a physical piece or pieces of whatever crashed in Mack Brazel's pasture.

2. evidence of the gouge (which would eliminate the balloon explanation).

We hope that Bill Doleman will follow through on this exciting find, as his opinion (as an independent expert) as to what it means is critical.

Tom

From: Landsman, Larry
Sent: Thursday, October 03, 2002
To: Carey, Tom; Schmitt, Don
Subject: Re: FW: One more thought

I've authorized Bill to go back in and complete his
analysis. In fact, we're sending MPH back to the site
to not only document Bill's reaction (startled), but to
document the anomaly. Cool, huh?

Larry

....

October 2, 2002
William Doleman, Ph.D.
Principal Investigator
University of New Mexico
Office of Contract Archeology
1717 Lomas NE
Albuquerque, NM 87131

Dear Bill,

This is to verify that we (SCI FI CHANNEL) are approving a 1.5-day
project extension as requested by you under section 3.B of the original
contract in a letter to me dated 9/27/02 (total cost $1,899). Further-
more, this letter also serves as authorization to invoice me for the SUP-
PLEMENTAL BUDGET for the Foster Ranch project technical and
logistical support which you outlined in your letter of 9/13/02 totaling
$2,831.

Please arrange to having UNM's accounting office either FAX,
email or snail mail me these two new invoices asap.

Thanks again for all the professional and thorough work.

Regards,

Larry Landsman
Director, Special Projects
SCI FI CHANNEL

....

From: Carey, Tom
Sent: Thursday October 03, 2002 10:48 AM
To: Landsman, Larry
Cc: Schmitt, Don
Subject: Re: FW: One more thought

Good deal. When we finally got back to Roswell after
the '79 Olds Cutlass broke down on the evening (late
evening) of Sept. 24, over a much needed beer at
Pepper's (the only place in Roswell still open), I told
Bill Doleman to write up his findings like he sees them—
not to shade anything one way or the other—for the sake
of all of our reputations. I don't know what the bagged
artifacts will prove out to be, but finding the "gouge"
was one of our "proofs" that what crashed in Brazel's
pasture was not a weather balloon. This is truly an
exciting development.

Tom

....

Wasting little time after receiving the official go-ahead, Doleman made several trips to the Foster Ranch site throughout October, and resumed his personal "Dig Diary" entries while performing soil tests and further investigating what he designated as an "alternative furrow." In addition to using a GPS nit to establish the furrow's exact location and a compass to determine its orientation, Doleman sketched out a scaled map to plot and measure the alignments of various furrow segments (copies of which he would later forward to Landsman).

October 5

On my way back from the New Mexico Geological Society's three-day field trip to White Sands Missile Range, I drop by the Foster Ranch site to take an additional look at the backhoe trench anomaly and the alternative furrow. It's pretty late in the afternoon by the time I arrive; I spend most of my time walking around the alternative furrow and looking for evidence that it's a road or some such explainable feature, but fail to find anything that connects it with the east or west the way a road ought to. It also doesn't run downhill the way one would expect an erosion channel to do, and it definitely lacks the "hump" you'd expect to see in the middle of an abandoned two-track road. A real puzzler. I take some preliminary notes, measurements and GPS readings, and make a "quickie" sketch map of the feature.

I continue on to the site and Study Unit 103's anomalous feature, which I've begun to call the "backhoe trench anomaly." By this time, the light is fading, and the feature's appearance is particularly disappointing. I leave it there and head for Albuquerque. I'll be back here with Bob and Lou next week.

October 9

Bob, Lou and I get a late start and head for the Foster Ranch site mid-morning. Our goal is to spend a couple of days, with Bob and Lou finishing the site map while I lose myself in the intricacies of soil geomorphology as displayed in the sides of the site's backhoe trenches. I also plan to investigate the backhoe trench anomaly and alternative furrow/weather balloon phenomena more.

We're joined by Alana Lynn Andrews, who had volunteered for September's main four-day excavation session and has gladly come back out to help some more. We meet at the Hines House and drive to the site, where Lou and Bob set up the transit and, with Lane acting as scribe, commence to work on the site map. I take off for the alternative furrow to take photographs of it and the weather balloon, and to collect the latter. I take photos of the alternative furrow from a variety of locations and directions with the goal of fully documenting its setting and unique characteristics, so others will see why I don't think it's a "normal" feature of the landscape. My opinion doesn't change at all.

I've also brought along an empty pinflag box, and a trash bag in which to place the weather balloon. After photographing the alternative furrow, I walk down to the weather balloon and take lots of photographs of it, including some that show it in relationship to the alternative furrow, and others that show it close-up, with the grass growing through it. Finally, I lift it off the ground, place it in the trash bag and put the whole package into the pinflag box. I then take photos of the ground left behind and the few scraps of balloon material that don't make it into the bag, which I leave as a final testament to where this incredibly ironic item has been found.

I also noticed a set of fresh tire tracks that went by the balloon on a northeast-southwest alignment and the east end of the alternative fur-row, where they turned west up the hill, paralleling the furrow. They are visible in one of the pictures I took of the alternative furrow (no. 1582). The fact that the vehicle tracks are manifest only as pressed-down grass suggests they're very recent—a few weeks old to a month or so, at best. As the balloon is considerably weathered and has at least one year's grass growth through it, and the mature yucca in the alter-native furrow attests to some age for the feature, it is clear that the ve-hicle tracks way postdate both phenomena, and thus cannot have been made by someone who planted the balloon or "manufactured" the al-ternative furrow. Nonetheless, the fact that the tracks go right by both is intriguing. A coincidence? Doubtful, but . . .

Time to head back to the motel in Corona (a lot closer than Roswell).

October 10

This afternoon we're joined by Melissa Jo Peltier (director), Miles Ghormley (sound man) and Bryan Duggan (cameraman) of MPH En-tertainment, the filmmakers who came back out from California to film the backhoe trench anomaly, the alternative furrow and weather balloon. Despite the fact that the filming slows everything down a lot, these people have been great to work with—good-natured and very professional.

Earlier in the morning, Lane, Lou and Bob continued working on the site map while I pressed on with my arcane soil-stratigraphy work in the trenches. I had never done soil-geomorph work in soils like these (limestone residuum), and was applying what I've learned from eolian setting to these super fine-grained deposits. Using HCl (acid) to monitor variations in calcium carbonate content was useless, as the soils are almost entirely derived from limestone, which is composed predominantly of—you guessed it—calcium carbonate!

With the MPH crowd arriving about the same time the crew fin-ishes up the site map, we eat lunch and head over to the alternative fur-row. Here Lou et al. begin making a measured sketch map of the alternative furrow, while the MPH folk film me talking about the fur-row and weather balloon. I had, of course, found the balloon two and a half weeks before, but they want to "re-create" the discovery, so I walk around until I "stumble" on it. I couldn't repeat what I actually said when I did find it, so I blurt out, "It's a freakin' weather balloon!" I can't believe I said that!

By the time MPH is happy with the alternative furrow/weather balloon footage, Lou, Bob and Lane are done with the sketch map. I have them reconnoiter the area around the alternative furrow to see if they can find anything else (a scrap of apparent weather balloon material was found about five meters south of the furrow, but nothing else), while Melissa, Bryan, Myles and I proceed to the backhoe trench anomaly for more filming. The sun is getting pretty low, and the anomaly is definitely not looking very good or obvious as it had on that fateful day of September 24. Oh, well. I talk and point, Melissa asks good questions and I talk some more. They get shots of me crouching in trenches and looking through my hand lens at dirt—doing the "science thing" (well, it *is* science, after all).

When all is said and done, we all take off for Corona, stopping at the Hines House and thanking Lane profusely for her help as we drop her off at her truck. I suggest we join up for dinner at the Willard Cafe, which is in the middle of nowhere—about 35 miles northwest of Corona—but which has some damn fine Mexican fare. We say our goodbyes and MPH heads to Albuquerque, while we head back to Corona for the night.

October 11

We had a little cleanup to complete at the site, so we head back out. I continue looking at dirt, while Lou and Bob pick up pinflags and backfill test pits, lining them with the plastic so they can easily be reopened in the future, should anyone care to do so. They also leave the rebars and gutter spikes marking the grid system centerline, so that future work at the site will not have to reestablish the system (a *lot* of work).

I actually begin to get a feel for the site's stratigraphy and complete documentation of the profile at SU 108, the backhoe trench across the electromagnetic conductivity anomaly, where I had found unusually deep soils and several large animal burrows that might explain the anomaly as being the result of moisture concentrated by either a bedrock depression, the burrow cluster or both. Although I've yet to complete my soil-stratigraphic studies, we head out for Albuquerque a bit early, as we had worked until late the day before. I'll just have to come back again.

October 16

On my own now. I leave OCA by 9 A.M. and drive directly to the Foster Ranch site to complete my soil-stratigraphy work, perform a final evaluation of the backhoe trench anomaly, and tie up a number of loose ends. On the way in, I take pictures of a true erosional channel in the bottom of a drainage and an abandoned two-track road to show why I think the alternative furrow is neither. I also get more shots of the alternative furrow, which begins to intrigue me more and more. What if there *was* a furrow? What if Don's eyewitnesses were right, but just one valley off? A second "skip"?

At the site, I start by taking pictures of all the cross-furrow backhoe trenches, including enough to cover all parts of all trenches. The sun angle requires that I do the north profiles so they'll be illuminated. When I get to SU 103, the one with the "V"-shaped anomaly, I take some more pictures to demonstrate that, despite some fading, it is still visible. I then proceed to take a slew of site shots to show the setting, and then pictures of the various rock features, including the cairns. I spend a lot of time trying to match up one of these with the picture of the "debris field" published in the 1994 Randle and Schmitt book *The Truth behind the UFO Crash at Roswell* (I think I eventually nail it at Cairn 2).

Finally, I hop into SU 109, the backhoe trench across the metal detection anomaly, and do a full soil-stratigraphy profile. This is the one with no Bk horizon, but lots of extra clay, which suggests a possible explanation for the anomaly—namely, higher ferromagnesian mineral content, including clays, all of which may be getting brought in by subsurface water flow, which may also be flushing the carbonates. Hhmmmhh . . .

After taking notes on several other backhoe trenches and test pits, I return to the SU 103 anomaly and—after being sure I've recorded all I could—put the feature to the test by scraping and cleaning the profile to see if the anomaly will still be there. If not, then it must have been a superficial artifact of the profile, despite its distinctive appearance to begin with. After scraping, it's pretty much gone.

Disappointed, I go off to photograph the CUFOS test pits. I then return to the SU 103 anomaly, clean it up a bit more, spray it to bring out any color differences and take one more photo. Although faint color differences are visible, they have changed location.

It's back to Corona for grilled hot dogs.

October 17

I drive back out to the site to do some more soil-stratigraphy and take more photos. I do full profiles for SUs 107 and 4 (soil-stratigraphy test pit), and take soil samples from the backhoe trench anomaly in SU 103 to see if the color differences could be verified through Munsell color determinations. I photograph a number of the archeological test pits for the record and some more CUFOS test pits, the coyote burrow near SU 8, as well as some more site overviews, until I run out of time and storage space on the digital camera card. Finally, I bid adieu to the Foster Ranch site and head back to Albuquerque, hoping I'd come back someday to test the backhoe trench anomaly and perhaps do more stripping of the surface sediments, in search of more HMUOs.

It's funny, but even the most mundane landscapes become a home away from home for me (and, I suspect, for many archeologists). Foster Ranch has become one of many fondly remembered, out-of-the-way New Mexico places about which I'd reminisce over future campfires. This certainly has been the most interesting—and most unusual—archeological project in my experience. But that's what makes an archeologist—we certainly don't do it for the money!

Hey, Larry:
Here's a copy of my "alternative furrow" notes and sketch map. As you can see, my handwriting leaves something to be desired. Hopefully, you can read this OK.
Bill

Notes on "Alternative furrow" (10·5·02)   Max depth 35 cm

Upper (W) end orient'n (fr W) = 100°

middle orient'n (fr W) = 80°   100° TN

Lower (E) end orient'n = 70°

GPS rd WPT 001 (0472144 E   1263 84 Yuccas
(±12.7 ft / EPE 14 ft) 3755555 N)

ca 13 min @ 5:15 PM

(→ NAD 27 = E 472 194
          N 3755351

Weather balloon ca 100 m   70° TN
not as big as I remembered

70 cm long by 20-30 wide

GPS rd WPT 002   ca 16 min @ 5:38 pm

       0472224 E
       3755512 N   ± 9.1 ft   EPE 10 ft

Go to WPT 001 = 288° 0.06 mi

Photo'd this year's grass that had grown into balloon

Local slope aspect est'd at 62½° (TN) from USGS map

So alt furrow ∡ differs from actual slope aspect by 37.5 (upper end)
     or by avg of 21.5° based   17.5° (central part)
     on orientation in LR's map (84°)   7.5° (lower part)

up slope cont'n of repressed but unevaded area

limit of current evasion

Upper sect = ca 6 m x 2.5-3 m

Central sect. 11 m x 2-2.5 m

Lower sect 15 m x 2 m

(not to scale)

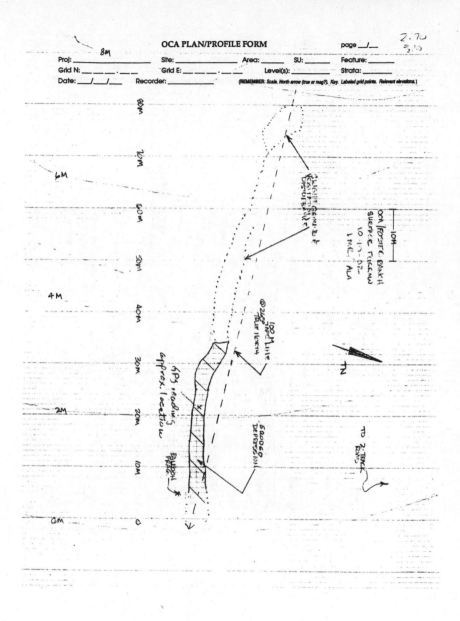

OCA PLAN/PROFILE FORM

page ___/___

Proj: _____   Site: _____   Area: _____   SU: _____   Feature: _____

Grid N: _____   Grid E: _____   Level(s): _____   Strata: _____

Date: ___/___/___   Recorder: _____   (REMEMBER: Scale. North arrow (true or mag?). Key. Labeled grid points. Relevant elevations.)

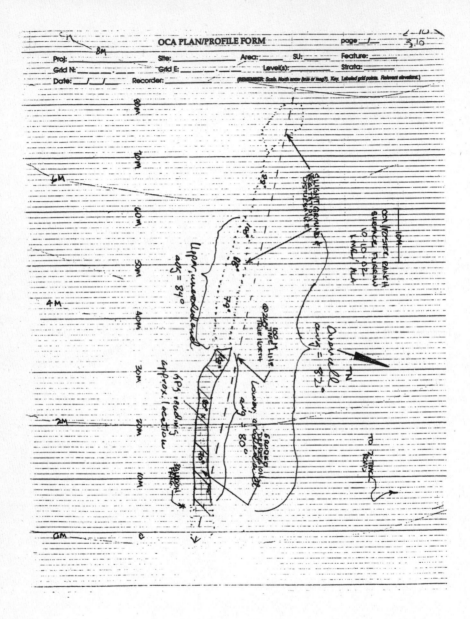

OCA PLAN/PROFILE FORM   page: 1

Proj: _____   Site: _____   Area: _____ SU: ____   Feature: _____
Grid N: _____   Grid E: _____   Level(s): _____   Strata: _____
Date: __/__/__   Recorder: _____   (REMEMBER: Scale. North arrow (true or mag?). Key. Labeled grid points. Relevant elevations.)

....

Not included among Doleman's diary entries (though taped for SCI FI's Roswell documentary) was an October 9 early morning visit to Doleman's office in Albuquerque from New Mexico gubernatorial candidate Bill Richardson. After reviewing the objectives and discoveries made at the alleged crash site, Richardson offered the OCA project director any services the governor's office could provide, stating in a postdiscussion interview, "I don't think the U.S. Government has fully disclosed everything they know. So it's important that, through science and through the Foster Ranch Project, we get to the bottom of something that's perplexed many Americans. Don't do it through theories or hearsay, but through serious *science*. This is very important. [The Roswell Incident] is a part of American culture . . . and I think through science, through serious people doing serious work, maybe we can learn something we didn't learn before. That, in itself, is valuable."

....

From: Doleman, Bill
Sent: Friday, October 18, 2002 5:26 PM
To: Schmitt, Don; Carey, Tom
Cc: Myself; Landsman, Larry
Subject: Re: The "balloon," the "alternative furrow" and the "anomaly"

Dear Tom & Don:

Sorry if you didn't get filled in completely about the two last-minute "finds." Things were pretty hectic that last day and have remained so since. I know Larry told Don about the "alternative furrow" and balloon, but I'm less certain how much Tom learned.

Anyway, to make a long story as short as possible, on the last day, when total havoc was in full swing, I found two interesting things. The first, located about 0.4 mi SE of the dig site was a 30-meter-long "furrow" on a hillside, and an apparent large balloon about 110 yds. to the SE. I have taken to calling the feature the "alternative furrow" and the apparent balloon a "weather balloon" in corresponding with Larry and MPH.

Extraordinary security measures were implemented to ensure the integrity of the project. At the conclusion of the primary phase of the dig on September 25, 2002, fifty-three bags of soil samples and twenty-four envelopes of HMUOs (Historic Materials of Uncertain Origin) recovered in the course of the excavations were transferred to lock boxes at the Wells Fargo Bank in Roswell.

The "furrow" is located about 100 yds. south of the 2-track that leads SW from the windmill, on the west side of the first little valley the road crosses; it is clearly visible from that road. I and my crew noticed it early on, but I had no opportunity to investigate it until the last day, while I awaited the slowly-moving backhoe. I parked at the top of the little valley's west ridge (so the backhoe could see my vehicle). From the ridge I could no longer see the "furrow," but I walked to the SE to where I thought it was. I overshot it and then, when I realized my error, I turned back to the north. While tracking back I found a weathered bundle of what appears obviously to be the remains of a large balloon. The upper portions that have been exposed to the sun are quite sun-dried and crumbly and shredded, while the lower, protected portions are more intact and still quite stretchy. The material appears to be latex or rubber of some kind.

The second thing of interest I found the last day was an apparent "anomaly" in one of the backhoe trenches we dug across the original furrow location, as determined by Don the first day (9/16) and again the last (9/24); I've sent you both pictures of that, too, I think (if not, let me know and I'll send some on). I found it late in the afternoon—as the excavations were winding down and the backhoe was digging—while frantically trying to catch up on my site photographs. I thought to myself, "Well, let's go see what the backhoe found." In the third trench down (just south of the little ridge, near the presumed north end of the furrow), I saw an apparent anomaly—that is, something that didn't fit with the natural stratigraphy. In cross-section, it basically looked like an asymmetrical V-shaped "channel" (with a rounded bottom) in the south trench wall. It was pretty much right where the evidence of a furrow or gouge would be if such existed. At first

On October 9, 2002, William Doleman briefed gubernatorial candidate Bill Richardson at the university. Doleman confirmed that he discovered a stratigraphic anomaly right where the gauge would have been the deepest. In response, Richardson said, "I don't think the U.S. government has fully disclosed everything they know. And what I think is important is that we, through science, through the Roswell Dig Project, get to the bottom of something that's perplexed many Americans for so many years."

glance, the bottom of the more gently sloping left side seemed to have a lens of clay deposits that might have formed as the feature filled in with water-borne deposits. I looked at it and literally said, "Holy shit! I've got an anomaly right where there should be one!" I took three pictures of it in the fading light, then ran off to do the "Tom-takes-Bill-to-the-site" re-enactment and my interview. That finished about 6:30, after which—as Tom knows—we went very slowly home. When I looked at the photos, it looked even more convincing, with soil color differences to the right and the middle (filled in part) looking slightly "caved-in," owing to the deposit being softer (just what you'd expect from a filled-in feature). I got pretty excited and sent Larry a copy; he got excited, too.

Based on these two finds, Larry authorized a project extension to cover the costs of investigating both of the above "finds," as well as finishing our site map and me doing a study of the site's natural stratigraphy. The stratigraphy study is necessary to provide a basis for interpreting any trench anomalies (like the one I found). Larry also had MPH come out again and file the two "finds." I actually ended up going back twice, once on Thur. & Fri., 10/10-11 (when half my time was lost to filming interviews and "science-in-action shots"), and again this week (Wed. & Thurs., 10/16-17, when I got some work done).

When I returned the first time, I looked at the trench anomaly again; it was much fainter, and some of the "clay lenses" had disappeared, even though I had not touched the feature. MPH filmed me delineating it with the tip of my trowel and talking about it and what I thought it might be. I also looked at the other (north) trench wall across from the anomaly and found no evidence of it at all (photographed it, too). On the final trip this week, it still didn't show up too well. I think two reasons it faded are:

(1) The sun was higher on 9/24 (earlier in the afternoon) than 10/10 (a bit later in the afternoon and 16 days later, with the sun not as high owing to the onset of Fall);

(2) The surface of the trench was still moist with natural soil moisture on 9/24, but had dried thoroughly by 10/10.

Finally, when I returned this week, I investigated the anomaly some by scraping 1-2 cm from the surface to get a better look at it and to confirm its existence. The outline almost disappeared completely, although I think it shows up in the photographs I took (not downloaded yet). I took several photos, including some where I tried to re-create the initial appearance by spraying the trench wall with a mist of water, and that's the photo in which I think it shows up.

When I first saw the anomaly, and as I thought about it between then and my return to the site, I tried to think of what could have created it. I came up with the following possibilities:

(1) The "gouge" we're looking for,
(2) A coyote burrow,
(3) A buried erosional channel,
(4) An artifact of the backhoe.

Here's my thinking:

(1) One would have to first eliminate "non-gouge" possibilities before tentatively concluding "gouge."

(2) Coyote burrow: in my many years of fieldwork, I have seen numerous coyote burrows, and there is even a recently abandoned one on-site (John Johnston was dying to dig it, but I dissuaded him because it wasn't nearly 55 years old). The coyote burrows I've seen angle downward into the earth, especially where there is no rock or steep slope to dig under or into. I photographed the one on-site and measured the entrance angle as well as the surface's slope angle. The burrow is about 2-3 feet deep and the entrance angle is 42 deg., while the surface's slope is 3 deg. (i.e., nearly flat). The gentler left side of the anomaly has a slope of ca. 30-35 deg., while the right side is ca. 40 deg. Could it be the cross-section of a collapsed burrow? Maybe.

(3) A buried erosional channel: As the feature does not appear on the opposite trench wall, it is likely not an erosion feature. Furthermore, its V-shaped profile is not consistent with most such channels. Finally, it lacks gravels in the bottom (quite common).

(4) An artifact of the backhoe: Since the anomaly nearly disappeared after just a little scraping of the trench wall, I have begun to think myself quite the fool—"fooled," that is, by the backhoe. That is, the backhoe created an "anomaly" a few cm deep in the face of the trench. The soil moisture content at the time of excavation was pretty high by desert standards, owing to the heavy precipitation of 9/18 and the relatively low evapotranspiration rates of late Summer/early Fall. Alley (the backhoe operator) is one of the best, but his bucket wobbles occasionally and creates phony "features" in the trench walls. The 30–35 deg. left side and steeper (40 deg.) right side of the anomaly are consistent with the direction of his digging (from left to right—shallow cut down, steeper pull up; I need to interview him about his excavation techniques, though). Also, the clay lens on the left side might have resulted from "smearing" of clays in the clay-rich sediments. Backhoe artifact? Quite possibly, I'm afraid.

Well, I hadn't quite planned to write so much, but now you know what I know and think. Please let me know your comments and if you need more pictures.

Cheers,
Bill
==========
William Doleman, Ph.D.
Principal Investigator
University of New Mexico
Office of Contract Archeology
1717 Lomas NE
Albuquerque, NM 87131
==========

....

From: Doleman, Bill
Sent: Tuesday, October 29, 2002 7:15 PM
To: Schmitt, Don; Carey, Tom
Cc: Myself; Landsman, Larry
Subject: "Balloon," "furrow" and "anomaly"—PART II

----------Forwarded Message----------

Dear All:

Well, I started this "followup" e-mail over a week ago,
but have been swamped with this and that. I'm writing
to you all so you know the results of my last visits to
the site to wrap up my investigation of the soils and
stratigraphy at the site, as well as what I learned
about the interesting anomalies that were found at the
last minute. This will all go into my report, so I'm
actually drafting my report at the same time!

I left a couple of things out of the first
"balloon/furrow/anomaly" e-mail. Also, I have an update
on the balloon's I.D. Here they are, including an
interesting "mystery" (how else am I gonna get you to
read such a long e-mail?):

THE ALTERNATIVE FURROW ("AF" for short; just my term
for it):

I neglected to discuss the "alternative furrow" and my
thoughts about its possible origin. As with the trench
anomaly, I approach this question by eliminating the
obvious possibilities. The feature's setting and
characteristics are important to evaluating these
possibilities. The AF consists of an elongated, eroded
"channel" that runs at an angle across the west slope
of a drainage southwest of the windmill. The eroded
area is flat-bottomed and about 2–2.5 meters across, 30
meters long and exhibits a fairly consistent width
except at its western upslope end, where it's rounded
off. This uphill edge and the feature's sides are
vertical, suggesting that the erosion process is
ongoing. At the eastern (downslope) end, the erosion
becomes increasingly shallow, eventually blending into
the natural, uneroded slope. The AF lies on the gently

sloping side of the drainage and is oriented at 15-45
deg. to the prevailing slope (i.e., it doesn't run
straight downhill like water and the erosion it creates
are supposed to). Another important consideration is
the general nature of rainfall-induced erosion in arid
settings as it occurs in various kinds of soil and
vegetation, both naturally and as the result of some
non-natural initiating factor (usually some disturbance
of the natural, vegetated surface).

As possible explanations, I came up with:

(1) Natural erosion channel: First, I surveyed the
project area for other examples of hillside erosion
channels and found none—I mean none. Fact of the matter
is, the local landscape is comprised of mostly gentle,
uneroded slopes and fairly dense grassland vegetation
resulting from an estimated avg. annual rainfall of
13 inches (compare with Albuquerque's 8 inches). In
other words, the most surfaces are stable and uneroded
areas, but they all occur in the bottoms of drainages
(e.g. the long one east of the Hines House), not on
the sloping sides where the AF occurs. Second, the
feature's alignment (it curves some) ranges from
70-100 degrees bearing from true north, while the
prevailing slope is estimated at 55 degrees bearing.
That means the AF runs at an angle of 15-45 degrees
to the slope and to the alignment any natural erosional
channel should have. This, together with the absence
of any naturally-occurring side-slope erosional
channels in the area, strongly suggest the feature
is not natural in origin. Thus, it must have been
induced by a non-natural disturbance of the surface,
which leads us to other possible explanations to be
considered:

(2) Abandoned old 2-track road that has since eroded:
as you all have seen for yourself in the project area—
particularly in the road we made ourselves—2-track
roads are created by driving back and forth across
undisturbed ground repeatedly until the 2-track is
produced. In 25 years of archeology in the boonies, I
have seen lots of them, and they all erode the same way:
two ruts and a hump in between continues to support

vegetation, at least until the ruts are so deep that
vehicles bottom out and a new 2-track is established
nearby. In a few rare cases, I have seen 2-tracks
become true arroyos in which the rut/hump topography is
completely erased, but this is clearly not the case
here. I even found and photographed an abandoned
2-track near the road to the windmill from the Hines
House. The ruts are beginning to be re-vegetated, but
the basic characteristics are still present: ruts and
hump. In addition, well-used 2-tracks tend to persist
for a long time and along their entire length. Aside
from slightly depressed and uneroded uphill extension
of the AF (see below), there is no evidence of any old
2-track segment in the middle of nowhere with no
connection to anything. I conclude it's NOT an old
2-track.

(3) For similar reasons, it does not appear to be an old
cowpath (too narrow, no evidence of its continuation at
either end). Also, there's a pair of active cowpaths
just to the south; these continue a long ways up slope
and down (all the way to the windmill to the east, in
fact).

(4) A non-natural feature produced by some localized,
and possibly elongated, disturbance of the original
ground surface sufficient to induce erosion. "Non-
natural" because of its non-downslope orientation and
location on a slope vs. a drainage bottom; "localized"
because there is no evidence of its continuation at
either end; "elongated" because of its dimensions and
parallel sides. Finally, when we returned to film the
AF, I had my crew prepare a standard map of it. They
were the ones who pointed out to me that the AF actually
extends uphill for another 30 meters as a shallow and
uneroded slight depression, with more or less straight
and gradually converging sides and a slightly concave
bottom. This extension is faint, but clearly
discernible as subtle topography and vegetation
differences. Essentially, the extension is just what
you might expect from something "grazing" the surface
and creating an elongated parallel-sided depression
that is tapered and shallowest at the ends. The eroded
portion is where you'd expect it to be, as erosion—or

"headcutting" as it's often called—generally starts at the bottom and works its way uphill.

So, there you have it. The AF remains an anomalous feature as best as I can tell, applying 25 years of field experience, common sense and specialized training in geomorphology and soils. It is consistent with the idea of a "furrow" as I understand it. I recently acquired the original aerial photography from 1946, 1954 and 1996 that Dave Hyndman analyzed as part of his contract with SCI FI. I have yet to look at them, but will see if the AF shows up on any of them. As soon as I have, I will forward my observations to you all.

THE BALLOON (definitely):

The "balloon" was found about 110 yards to the southeast by complete accident when I first went to check out the "furrow." As I noted in the first e-mail, the part that has been exposed to the sun is quite weathered and brittle, while the lower protected part is still elastic, but tears easily. I took it to the Nat'l Weather Service office here, and they confirmed that it certainly could be one of theirs, but noted that other agencies, including several Dept. of Defense outfits (Kirtland, LANL, White Sands) use high-altitude payload balloons for a variety of experimental purposes. The specimen is partly shredded, a condition consistent with what happens to weather balloons and other high-altitude payload balloons when they reach great heights and expand to the point where their elastic limit is exceeded and they burst (going from ca. 6 ft. in diameter to house size). They showed me one of the ones currently in use, and the materials are quite similar. They hazarded a guess that it probably "isn't more than 10 years old," but expressed lots of uncertainty. They thought maybe the manufacturer (currently Kaysam, Inc.) might be able to better date it. They also confirmed that it had undoubtedly been released somewhere in the west, as its final demise would have occurred way high where the winds would blow west to east. Finally, when I collected the balloon, I had some difficulty removing it because there was grass

growing through the shredded material, indicating it
had been there at least a year of so.

THE MYSTERY:

A single set of recent vehicle tracks goes right by the
balloon and the alternative furrow. I noticed the
tracks when I first found both. They're not mine, and
although MPH drove in to the AF with their equipment,
the tracks pre-date that episode. From past experience,
the "mystery tracks" look to represent a single pass
that is less than a year old. I couldn't tell the direct
driven, but the tracks go past the balloon on the east
(downhill) side and northwest to the AF, again on the
east side, then turn southwest and go up the hill
paralleling the AF on its north side. Coincidence? Hard
to imagine. Did someone plant the balloon and create
the AF? I think the balloon has been there at least a
year longer than the tracks, and the AF's erosion has
been in the making some time.

That's all, folks!

Cheers,
Bill
==========
William Doleman, Ph.D.
Principal Investigator
University of New Mexico
Office of Contract Archeology
1717 Lomas NE
Albuquerque, NM 87131
==========

....

By late October, SCI FI had finally distributed a press release announcing
its excavation of the Foster Ranch, the results of which would be revealed in
a two-hour prime-time documentary November 22. Although Doleman—
having just returned to the University of New Mexico—wouldn't have all of
the massive data compiled by that time, the network already had what it
needed for the special, including details and discoveries made at the Foster
Ranch excavation, new eyewitness testimonies and "smoking gun" evi-

dence—a computer-enhanced enlargement of a July 1947 photograph that would enable investigators to "read" a telex in General Roger Ramey's hand.

FOR IMMEDIATE RELEASE

SCI FI CHANNEL SPONSORS LANDMARK
ARCHEOLOGICAL EXCAVATION
AT 1947 ROSWELL CRASH SITE
Groundbreaking Scientific Investigation
to Be Chronicled in SCI FI Documentary
"The Roswell Crash: Startling New Evidence"
Hosted by Bryant Gumbel on November 22

New York, N.Y., October 29, 2002—As part of SCI FI Channel's recently announced advocacy initiative to help bring scientific, congressional and media attention to the UFO phenomenon, the Channel turns to the tools of modern science to help unravel the decades-old mystery of the "Roswell Incident." Utilizing state-of-the-art remote sensing technologies and modern archeological forensic science under the supervision of the University of New Mexico, SCI FI announces its coordination and sponsorship of a landmark scientific excavation of the 1947 crash site. Working under top secret conditions, skilled archeologists set out to unearth conclusive physical evidence to help prove or disprove what some claim is science fiction—the crash of an extraterrestrial craft.

Considered by many as the "Holy Grail" of all UFO stories, the "Roswell Incident" has captured the imagination of the public for years. Did an actual UFO crash outside of Roswell, N.M., in July 1947? Out of those Americans who know of the "Roswell Incident," less than three in ten (28%) believe the "official" government story of a weather balloon crash at that site, according to a 2002 national Roper poll commissioned by SCI FI.

To chronicle this groundbreaking archeological investigation, SCI FI sent its documentary cameras into the deserts of New Mexico for **THE ROSWELL CRASH: STARTLING NEW EVIDENCE,** hosted by **Bryant Gumbel** (CBS's *The Early Show,* NBC's *Today Show*). Premiering on Friday, November 22 as part of a full night of special programming beginning at 8PM (ET/PT), this new two-hour SCI FI documentary of the "Roswell Incident" includes all-new eyewitness interviews and up-to-the-minute late-breaking revelations.

From the initial headlines of a "disk" being recovered in the desert in 1947 to SCI FI's latest "smoking gun" bombshell, this new examination of the "Roswell Incident" offers the definitive account of what may be the most important event of the modern age.

THE ROSWELL CRASH: STARTLING NEW EVIDENCE is directed and executive produced by Melissa Jo Peltier of MPH Entertainment (*My Big Fat Greek Wedding, The Lost Dinosaurs of Egypt*). James Romanovich of Platinum Media, Inc. also serves as executive producer.

In support of **THE ROSWELL CRASH: STARTLING NEW EVIDENCE,** SCI FI Channel's award-winning website SCIFI.COM, will offer exclusive background resources on this historic archeological project, as well as the "Roswell Incident" itself (users can log on directly through http://www.SCIFI.COM/UFO):

- SCI FI LIVE CHAT—Wednesday, October 30 at 9PM (ET)/6PM (PT) UNM's lead principal archeologist **William Doleman, Ph.D.** will discuss aspects of the top secret excavation.
- SCI FI ROSWELL DIG DIARY—Starting November 13, a complete day-to-day diary that documents the entire groundbreaking ten-day event will be posted. These daily entries will detail the project from various points of view including those of **Doleman,** veteran Roswell investigators **Tom Carey** and **Don Schmitt;** and SCI FI Channel's senior vice president of programming, **Thomas Vitale,** among others.
- SCI FI LIVE CHAT—Friday, November 22 at 10PM (ET)/7PM (PT) **Carey** and **Schmitt** will discuss the investigation of the "Roswell Incident" itself.
- THE ROSWELL REPORT—Dedicated to informing those who desire to know the truth behind an extraordinary event that occurred more than 50 years ago, this ongoing column offers up new information stemming from **Carey** and **Schmitt's** still-continuing, intensive investigation into this remarkable case of apparent extraterrestrial visitation.

THE ROSWELL CRASH: STARTLING NEW EVIDENCE is an MPH Entertainment production for SCI FI. Launched in 1996, MPH has produced over 130 hours of primetime television programming and two independent feature films, including co-producing the 2002 smash hit *My Big Fat Greek Wedding*. Notable among MPH's many television projects are cable's *The Lost Dinosaurs of Egypt, Founding Fathers,* Discovery Channel's *Eco-Challenge Australia,* SCI FI Channel's *Martian Mania: The True Story of the War of the Worlds, Las Vegas: Gamble in the Desert* and *Sea Tales.*

The national opinion poll on The Roswell Incident was conducted by **RoperASW** among a representative sample of 1,033 adults ages 18 and over. The telephone interviews were conducted from September 6th through September 8th, 2002. The margin of error for the total sample is +/- 3%.

SCI FI Channel transmits fantastic images to 79 million human homes. Launched in 1992, SCI FI features a continuous stream of cinematic hits, new and original series, and special events, as well as classic sci-fi, fantasy, and horror programming. Check out SCIFI.COM®, the SCI FI Channel's award-winning website, at *www.scifi.com*. SCI FI Channel is a program service of Universal Television (*www.univer-salstudios.com*), a division of Vivendi UNIVERSAL Entertainment (VUE), the U.S.-based film, television and recreation entity of Vivendi Universal, a global media and communications company.

....

Prior to the special, SCI FI received a final report from Sunbelt Geophysics' David Hyndman and Sidney Brandwein. Based on ground conductivity and metal detection surveys of the site, plus examination of pre- and postcrash aerial photography, their geophysical investigation—the results of which are summarized in Bill Doleman's final report—was unable to offer conclusive evidence that a UFO had crashed around the suspected impact site in 1947. But neither could it rule out the possibility; only a small portion of the ranch's vast landscape had been excavated, and the existing anomalies discovered in that area alone clearly supported the supposition that further research, including a more extensive excavation, was warranted.

The amount of viewers tuning to the SCI FI Channel on November 22, 2002, would likely support such an assessment. *The Roswell Crash: Startling New Evidence* unearthed the network's highest ratings ever for an original special, validating several key points in the case while bringing to light new questions, the most compelling among them concerning the "HMUOs" and soil anomalies that had been uncovered and secured in Roswell's Wells Fargo Bank.

....

November 25, 2002
SCI FI DIGS UP RECORD-BREAKING RATINGS
Roswell Documentary Becomes Highest-Rated
Original Special in SCI FI Channel History

On *Friday, November 22,* in anticipation of its upcoming epic 10-night miniseries, *Steven Spielberg Presents TAKEN,* SCI FI hosted a night of original documentary specials aimed at unearthing the truth behind UFOs, government conspiracies and alien abduction. In the process, the Channel uncovered a ratings winner. SCI FI was **the #1 adult-targeted cable network from 8–11 PM with a 1.7/1,342,000 HHs***.

Intrigued by SCI FI's smoking gun shocker, **an average of 2,365,000 viewers**** tuned in for *The Roswell Crash: Startling New Evidence,* hosted by Bryant Gumbel. The documentary, chronicling the Channel's landmark scientific excavation of the 1947 crash site in Roswell, New Mexico **garnered a 2.0 rating (1,560,000 HHs)***— making it the highest-rated original special in SCI FI Channel history.** *The Roswell Crash* surpassed the previous record-holder, the one-hour special *Curse of the Blair Witch* (1.8 / 1,032,000 HHs 7/12/99), by over half a million HHs.

Following *The Roswell Crash,* the 10 PM documentary *Abduction Diaries* **earned a strong 1.1 (901,000 HHs)*****, with **the overall primetime block earning a 1.7 (1,340,000 HHs)***—trying with the SCI FI original movie *Sabretooth* for the highest-rated night this November.

---

* source: NMR/Galaxy, Friday 11/22/02, 8–11PM coverage area time period rating
** source: NMR/Galaxy, Friday 11/22/02, 8–11PM, average coverage area time period HH rating/delivery
*** source: *Nielsen Media Research/Galaxy Explorer, Friday 11/22/02, 8–10 PM coverage area rating
**** source: NMR/Galaxy, Friday 11/22/02, 10–11PM coverage area rating

TAKEN, the SCI FI Channel's ambitious science fiction adventure that weaves together over 50 years of alien abductions into the compelling story of three families' experiences, premieres *Monday, December 2, 2002 @ 9pm ET/PT.*

SCI FI Channel transmits fantastic images to 80 million human homes. Launched in 1992, SCI FI features a continuous stream of cinematic hits, new and original series, and special events, as well as classic sci-fi, fantasy, and horror programming. Check out SCIFI.COM®, the SCI FI Channel's award-winning website, at www.scifi.com. SCI FI Channel is a program service of Universal Television Group (www.universalstudios.com), a division of Vivendi UNIVERSAL Entertainment (VUE), the U.S.-based film, television and recreation entity of Vivendi Universal, a global media and communications company.

....

At this point, Bill Doleman was well under way in trying to answer those questions, correlating his own extensive notes with those of his assistants, J. Robert Estes and Louis Romero, reviewing the Sunbelt Geophysics report and deciphering the volunteers' grid excavation records, which documented all of the test pits that were excavated at ten-centimeter levels. Soil-stratigraphy records, meanwhile, were not only crucial in helping Doleman evaluate the backhoe trench anomaly but enabled him to hypothesize the origins of other anomalies.

The process was not without setbacks. When the archeologist visited the Wells Fargo Bank in April 2003 to examine the stored excavation materials, he discovered that latent moisture, trapped inside the double-boxing of soil samples, had caused the deterioration and partial or total disintegration of approximately one-quarter of the paper bags containing the samples. Fortunately, enough materials remained intact for Doleman to continue inspecting them, and to correspond his findings to Landsman, Don Schmitt, and Tom Carey.

....

From: Bill Doleman
Sent: Monday, June 02, 2003 8:52 PM
To: Landsman, Larry
Cc: Schmitt, Don; Carey, Tom; Myself
Subject: ROSWELL SCIENCE—update

Dear Larry, Don & Tom:

Here's what I got so far on the various analyses. As I have noted in earlier conversations, there are two areas in which we need to have scientific analyses performed. The first is the HMUOs, and the second is the soil samples. The kinds of analyses required are generally quite different. I have made progress in both directions and been stymied some as well in one.

HMUOs

My "eyeball" analysis of the HMUOs led to pretty certain ID of most, but there are some whose nature I have guessed at, but cannot be certain about. These (5-10) need to be looked at by other scientists. They can be grouped into two categories:

(1) Apparent man-made items, including the "orange blob" (looks plasticey, but is weird), Chuck and Debbie's little gray thingy that looks vaguely like a scrap of trash bag plastic (thin, flexible, somewhat fragile), a piece of what looks like a white plastic pipe flared at one end, and another possible plastic fragment, and several boot/shoe sole fragments and some apparent leather (these, as Don has pointed out, might be datable and perhaps identifiable as to origin and use [i.e., possible by the military?]).

(2) Apparent organic thingies of biological origin, including the "alien condoms" that I think might be empty (i.e., formerly occupied) reptile eggs, as well as the weird green flat thingies that look like they might be some sort of dried algae mat, or . . . ?

It would also be nice to have the weather balloon positively identified and, if possible, dated. I must emphasize that the guy at the NWS in Albuquerque acknowledged that his 10-years-old estimate was just a guess.

I followed the recommendation from Pat Flanary and finally managed to get a hold of Mr. Ed Roberson at Roswell BLM, and he in turn contacted the folks in their criminal division. The upshot is that they are not comfortable approaching the FBI because no crime is involved (unless my personal theory of it being a single-vehicle WI crash is correct—hey, July 4th, New Mexico, nighttime—happens all the time!). They are afraid someone would ask why FBI resources are being wasted when there are so many unsolved federal crimes.

The big question remains what kind of lab and/or analyses need be done on the HMUOs. I believe the first thing to do is to have a forensic lab I.D. them, if possible. If I.D. is not possible, then we go to materials-testing and or chemistry labs to see if there's anything weird in them (see soil sample discussion below). This assumes—quite possibly erroneously—that the right forensic lab can I.D. anything, a notion I've no doubt gotten from watching too many "forensic" TV shows ("Crossing Jordan," the "CSI" and "Law and Order" series, etc.). The idea is that they maintain "libraries" of "thingies," both man-made (like the libraries of clothing and carpet fibers we've all seen them use) and natural (like the bug library maintained by "Bug" on "Crossing Jordan"). Actually, there are specialists archeologists use who maintain libraries of charred plants and various plants' pollen for I.D.'ing in archeological setting, so the concept is not unrealistic.

Forensic labs also rely frequently on various chemical analyses (of which there are a zillion types and methods, I'm learning), and such might be required for our non-natural HMUOs.

The important point here is that the right analysis(es) and the right lab(s) aren't as easy to pick out as one might think, and multiple, sequenced analyses may be required.

SOIL SAMPLES

I had a long chat with the chair of UNM's Earth and
Planetary Sciences Dept., who is a good friend and a
well-known geomorphologist specializing in the study of
desert soils, and he recommended two kinds of analysis
for the soil samples. The first would be a determination
of both common and rare elements present in them, the
idea being that if microscopic debris is present, it
should show up as anomalous amount of rare elements. He
gave me the name of the fellow here at UNM who does
X-ray fluorescence analysis (XRF). I contacted this guy
and he gave me a great deal of information about the
methods available, including ICP-MS (Inductively
Coupled Plasma-Mass Spectrometry), which can detect a
much wider array of elements. Searching for a wide
array seems reasonable, as I assume we would expect the
debris to be composed of unusual materials containing
unusual and possibly rare elements. The XRF guy also
gave me the names of two private labs that do such
analyses and one of these also does a slew of different
"forensic" analyses, and thus may be worth pursuing for
the HMUO end of things (I'd like to see what the guv can
do for us first, if that's still an option).

The dept. chair also suggested doing X-ray diffraction
studies to ascertain the mineralogy of the soils and
gave me another fellow to contact here at UNM, who
specializes in soils and clay. This fellow said "soils
are ugly" (for technical reasons) and suggested another
guy at UNM, and I left a phone message with him. X-ray
diffraction determines mineralogy by shooting X-rays
into specimens and analyzing the diffraction patterns
that come out the other end (X-ray diffraction was the
method used to first determine the helical structure of
DNA, a story recently detailed in a book about the
woman who did it, but got no recognition until after
she died). The mineralogy of the soil samples is of
interest because it may help confirm or deny the
hypothesis that Dr. Hyndman and I cooked up to explain
the metal-detection anomaly where we focused our
excavation efforts on the last two days—namely that
underground water flow (technically "vadose zone flow")
is concentrating iron/manganese-rich minerals dissolved

over the years from the limestone bedrock. According to
Hyndman, his metal-detection equipment is sensitive to
such minerals, as well as to pure metal.

Well, there you have it as it stands right now.

Cheers,
Bill
==========
William Doleman, Ph.D.
Principal Investigator
University of New Mexico
Office of Contract Archeology
1717 Lomas NE
Albuquerque, NM 87131
==========

....

By late July 2003, Doleman had submitted OCA's report of the J. B. Foster
Ranch project to the SCI FI Channel. His findings, coupled with the lab
analyses of the HMUOs and soil samples, would be unveiled October 24,
2003, in a follow-up of the Roswell dig on SCI FI's next *Declassified* special,
The New Roswell: Kecksburg Exposed.

....

FINAL REPORT
Archeological Testing and Geophysical Prospection
at the Reported Foster Ranch UFO Impact Site,
Lincoln County, New Mexico
WILLIAM H. DOLEMAN
OCA/UNM Report No. 185–769

INTRODUCTION

From September 16–24, October 10–11 and October 15–17, 2002, archeologists from the University of New Mexico (UNM) Office of Contract Archeology (OCA) conducted archeological testing and related research at the reported location of a low-angle extraterrestrial vessel impact in 1947 on the Foster Ranch in Lincoln County, New Mexico. This event—known generally as the "Roswell Incident," or the "Roswell Crash"—is one of the most famous UFO reports in the world, and as such, one of the most controversial. The research was privately funded by USA Cable's subsidiary, the SCI FI Channel, with Larry Landsman, Special Projects Director, acting as SCI FI's contact. Technical advisors to the project were Donald Schmitt and Thomas Carey, well-known independent UFO researchers. William Doleman of OCA served as principal investigator. The project, as well as the background of the Roswell Incident and new evidence derived from a 1947 photograph, was documented in *The Roswell Crash: Startling New Evidence,* first broadcast by the SCI FI Channel on November 22, 2002.

Archeological excavations were accomplished by volunteer excavators under the supervision of Doleman and OCA staff members J. Robert Estes and Louis Romero. In addition, geophysical prospection surveys (electromagnetic conductivity and metal detection) were conducted by Dave Hyndman of Sunbelt Geophysics in Albuquerque for the purposes of locating subsurface anomalies warranting archeological investigation. Finally, backhoe trenches were excavated across the reported location of a "furrow" reportedly created by the vessel impact and in two geophysical anomalies identified by the geophysical prospection.

Heretofore, evidence of the Roswell Incident has been limited largely to early newspaper reports, eyewitness accounts, signed affidavits, ancillary documents and interviews—many conducted long after the event itself. Much of this evidence has been summarized and critiqued in numerous articles and books, not to mention being an ongoing subject of discussion and dispute at conventions, and in Internet newsgroups, chat rooms and mailing lists. Many of those who were first involved as eyewitnesses or investigators are now dead. Throughout the 55-plus years that have ensued since the Summer of 1947, no evidence other than written declarations, recorded verbal testimony, and theorizing (fact-based and otherwise) has been offered to support claims that something—many believe an extraterrestrial vessel—crashed on the high-desert grasslands of southeastern New Mexico in 1947. In other words, despite reported accounts of one or more crashed vessels, non-human bodies, impact marks on the ground and unusual crash debris, none of this reported physical evidence has survived to the present. Or, if it has, it has not been made available for public scrutiny. Thus, the importance of the project reported herein lies in the fact that it represents the first-ever comprehensive attempt to discover physical evidence of the event, and to apply modern scientific methods to the search.

The area investigated lies on the former J. B. Foster Ranch approximately 48 kilometers (30 miles) southeast of Corona, New Mexico (Figure 1). The area chosen for the investigation represents the generally agreed-upon location at which the ranch's foreman at the time, William "Mack" Brazel, discovered unusual "debris" on an early July morning in 1947. This site is commonly known as the "debris field" and/or "skip site;" even the U.S. Air Force has acknowledged the site as the location where something came down in 1947 (albeit a Project Mogul balloon train, and not a UFO [Thomas Carey, personal communication, July 28, 2003]). For the purposes of this report, the term "Foster Ranch site" will be used. According to various testimony-based reconstructions of events, this location was the initial impact point of an object that hit the ground with a glancing impact, then rose into the air and crashed again some 25–40 kilometers (15–20 miles) away. The specific area subjected to intensive study during the present project measures 300 x 120 meters (984 x 394 feet), and comprises an area of 3.6 hectares (8.9 acres).

The land where the investigation took place is owned and administered by the United States Department of Interior, Bureau of Land Management, Roswell Field Office (BLM-RFO). Pursuant to BLM regulations, the work was conducted under Cultural Resource Use Permit No. 05-8152-02-01, which was issued by BLM-FFO upon approval of a testing plan submitted to BLM by OCA in early September, 2002. Patrick Flanary of the BLM-FFO acted as the BLM's contact for the project. It is also standard BLM procedure to withhold the exact locations of cultural resources in order to protect them from casual visitation by non-professionals. Site location information has been determined to be exempt from Freedom of Information Act inquiries, and the specific location of this project is being withheld until further notice.

The possibility of using archeological and geophysical methods to investigate the Foster Ranch site was first broached with OCA personnel by Mssrs. Schmitt and Carey in 1999. Having secured a commitment for funding from SCI FI in the Spring of 2002, Schmitt and Carey, together with Mr. Landsman of SCI FI, met with OCA Director Richard Chapman and OCA principal investigator Doleman in June to discuss appropriate research methods. The basic components of the research reported herein, including the use of subsurface geophysical prospection, were hammered out in that meeting. The specific geophysical prospection technologies used were determined by Dr. Hyndman.

The use of archeological methods is entirely appropriate to the search for physical evidence of whatever happened on the Foster Ranch in 1947 because archeology is essentially the forensic science of the past. This is true because the primary goal of archeologists is to find and interpret physical evidence of past events, and archeological methods are designed to do just that. These methods include modeling the dynamic processes that create the static archeological "sites" we see today, as well as those processes that subsequently modify them. Together, these processes are referred to as "archeological formation processes." In addition, geophysical prospection methods and technology—originally developed for detecting the subsurface presence of geological phenomena—were incorporated into the archeological fieldwork as an aid in choosing the locations of archeological explorations, and were warranted by the nature of the physical evidence being sought.

As noted, UNM's Office of Contract Archeology became involved when the SCI FI Channel agreed to fund the project. By far, most of OCA's contracts involve a variety of archeological fieldwork and report writing related to the identification, evaluation, and sometimes testing and/or excavation of archeological sites. Such projects are commonly occasioned by the need for federal or state agencies, or private institutions and commercial enterprises to comply with federal and/or state laws governing the treatment of cultural resources that may be affected by ground-disturbing activities. The term *cultural resources* refers to places with cultural or historic significance, and most commonly includes archeological sites, historic buildings and traditionally sacred locales. Foremost among federal cultural resource laws is the National Historic Preservation Act of 1966, while the primary state-level law in New Mexico is the New Mexico Cultural Properties Act of 1969, as amended in 1978.

These laws generally specify that any activities that may disturb cultural resources be preceded by archeological and historical surveys, to identify and evaluate the cultural and scientific significance of any cultural resources present, with "significance" being determined by specific criteria that define the eligibility of any resource to be placed on the National Register of Historic Places. Following survey, a reasonable attempt must be made to avoid adverse impact to any potentially eligible resources, by either avoiding them or recovering the valuable information they contain through excavation or some other form of appropriate documentation. In cases where the extent, significance or even presence of a site cannot be easily determined from surface evidence, archeological testing is often implemented as a means of learning more about the site. The Foster Ranch site project represents an example of archeological testing.

Although radically different from most archeological sites, the Foster Ranch site is amenable to archeological testing for several reasons. The first, mentioned above, is that the 1947 event reportedly produced physical evidence. The second is that, despite the controversial nature of both the site and the event, the site is of considerable cultural importance to the people of New Mexico, and even the entire world— more than one movie and a successful television series has been based on the Roswell Incident. Given this, the site might even qualify as a *traditional cultural property,* a term usually reserved for places of religious and cultural significance to Native Americans. Finally, the use of scientific methods to search for physical evidence of whatever

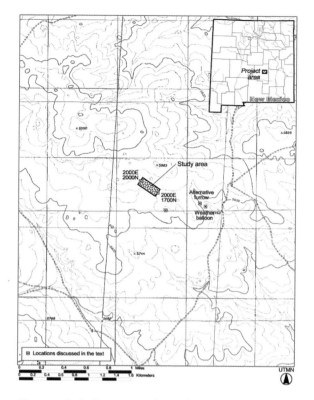

Figure 1. J. B. Foster Ranch site location and topographic setting. Note southeastward trend of study area drainages.

Figure 2. OCA staff member J. Robert Estes using a transit to map the Foster Ranch site, view northwest from south of Study Unit 8. Line of pinflags to right of two-track marks reported furrow centerline.

Figure 3. Volunteer excavators Debbie Ziegelmeyer and Chuck Zukowski at Study Unit 4, view northwest. Note screen and plastic washtub for collecting soil samples after screening.

Figure 4. A preexcavation photograph of Study Unit 2 surface.

Figure 5. An end-of-day photograph of Study Unit 2 after excavation on one level.

Figure 6. Excavation of Study Unit 103 by Elgio Aragon of Alleycat Excavating, with Study Units 101 and 102 in background, view north-northeast. Line of pinflags to left of two-track marks reported furrow centerline.

Figure 7. Typical karst topography at the Foster Ranch site, view east; light areas are bedrock outcrops. Note complete sinkhole beyond Dave Hyndman of Sunbelt Geophysics and crew establishing the east side of the grid system.

Figure 8. Foster Ranch site overview from Study Unit 6 (in foreground), view north. Vehicles and people are in swale; truncated sinkhole to left.

Figure 9. View south along grid system centerline from north end of grid system, Study Units 101–3, and Study Unit 104 being excavated.

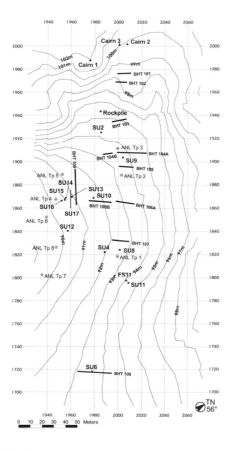

Figure 10. Transit-based topographic map of site grid system, showing excavated study units ("SU"), backhoe trenches ("BHT"), documented rock features ("cairn," "rockpile"), CUFOS test pits ("ANL Tp"), and one surface find ("FS 31," see text). Permanent rebar markers were established at the north and south ends of the grid system, and large nails mark the centerline.

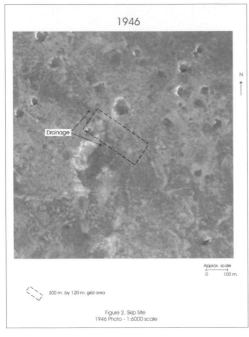

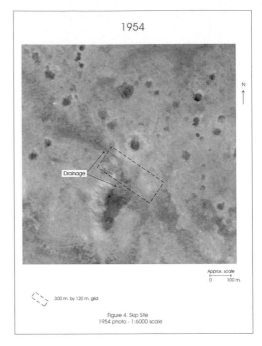

Figure 11. Enlarged portion of a 1946 aerial photograph showing the grid system outline and drainage traces. Circular features are natural sinkholes.

Figure 12. Enlarged portion of 1954 aerial photograph showing the grid system outline and drainage traces.

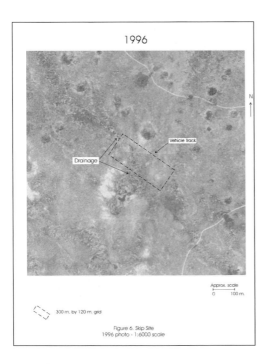

Figure 13. Enlarged portion of 1996 aerial photograph showing the grid system outline, drainage traces, and old two-track road (no longer visible on ground).

Figure 14. View northwest across synclinally tilted bedrock sections in northeast part of study area. Note northwest-tilted bedrock in foreground, and southeast-tilted bedrock in middle distance.

Figure 15. Computer-generated color-shaded contour map of electromagnetic conductivity survey readings. Low readings are blue, intermediate readings are green-yellow, and high readings are red.

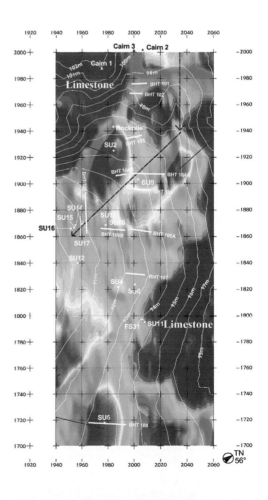

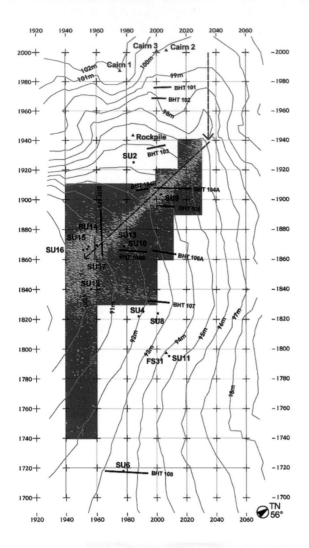

Figure 16. Computer-generated map of high-resolution metal detection survey readings. Low readings are blue, intermediate ones are green-yellow, and high ones are red. Note main elliptical anomaly; drainage-aligned linear trace entering from northeast and fainter trace from southeast are both drainages.

Figure 17. OCA staff member J. Robert Estes supervises volunteers Alana Andrews and Jerry Lowe at Study Unit 2, view west. Note numerous surface rocks indicative of shallow soils over bedrock.

Figure 18. Volunteers Nancy Easley Johnston and John Johnston at Study Unit 8 in bioturbated area, view northeast. Note concentration of disturbed soil and lack of vegetation.

Figure 19. Volunteers Chuck Zukowski and Debbie Ziegelmeyer at Study Unit 12 in drainage area, view east-northeast. Nancy Easley Johnston and John Johnston at Study Unit 10 in middle distance, Alana Andrews at Study Unit 9 in far distance.

Figure 20. Volunteer Alana Andrews at Study Unit 9, view northeast. The synclinally tilted bedrock feature—or "mini-syncline"—can be seen in upper left.

Figure 21. Volunteer Kate Leinster at Study Unit 11, view south. Note numerous surface rocks indicative of shallow soils over bedrock.

Figure 22. Volunteers Nancy Easley Johnston, Kate Leinster, Chuck Zukowski, and Debbie Ziegelmeyer at Study Unit 14 in the main metal detection anomaly, view east. Study Unit 13 is behind Mrs. Johnston; Study Unit 10 is in the left distance.

Figure 23. Study Unit 17 stripping trenches across main metal detection survey anomaly, view south-southwest. Pinflags mark Study Units 13–16.

Figure 24. Study Unit 108 south profile. Note darker A and Bw horizons above 40–45 cm, and lighter-colored Bk horizon from 45–90 cm downward. Also note two large, in-filled animal burrows at 70–100 cm, with right burrow showing evidence of stratified fill.

Figure 25. Study Unit 109, east profile, left side wetted. Note absence of well-developed, light-colored Bk horizon. Slight darkening in bottom 30–40 cm of profile represents increased clay content.

Figure 26. V-shaped stratigraphic anomaly in Study Unit 103 profile, taken September 24, 2002, at 3:13 P.M., about an hour after excavation. Note thin sediment line rising to left, color differences to right, and "caved-in" appearance of fill.

Figure 27. Anomaly in wetted Study Unit 103 profile, taken 4:44 P.M. on October 16, 2002, 22 days after discovery. Note change in bottom and right outline and partial persistence of thin sediment line defining left side, as well as color differences to right, and "caved-in" appearance.

Figure 28. Study Unit 103 anomaly profile, taken after scraping and rewetting at 6:22 P.M., October 16, 2002, almost two hours after Figure 27. Note near-absence of anomaly and shift to left in barely detectable color differences.

Figure 29. Cairn 2 in north end of study area, 7.3 meters east of 2000N/2000E, view 9 deg. west of grid north. Rock sizes, count, and setting match photo in Randle and Schmitt (1994).

Figure 30. Probable 1989 CUFOS test pit ("ANL Tp 1"), located near Study Unit 8, along the furrow centerline.

Figure 31. View southwest (218 deg. from true north) of the alternative furrow from two-track road into study area. Note how feature crosses valley side slope at an angle, and faint uphill extension to right. Faint linear feature at left edge of photo is a cowpath and not part of the alternative furrow.

Figure 32. Main portion of alternative furrow, view southwest (235 deg. from true north). Note the flat bottom, vertically eroded sides, and mature yucca growing in the upper part.

Figure 33. Main portion of alternative furrow, view east (80 deg. from true north). Note the curvature, mature yucca growing in foreground, and valley in distance.

Figure 34. Close-up of weather balloon found 100 meters southeast (true north) of the alternative furrow (divisions on scale are 10 cm). Note shredded and weathered condition and grass growing through fabric.

happened at the site must be deemed a laudable goal because only confirmed physical evidence—either of an extraterrestrial crash or some alternative explanation—will ever set the Roswell question to rest.

The research strategy and methods used in the Foster Ranch site investigation are described in the following section. The core of this research approach is based on the fact that, at present, no obvious evidence of either debris or a furrow is visible on the ground surface at the Foster Ranch site. Hence, the approach involved combining reports of the physical evidence once present and the alleged removal of most of it, together with knowledge of the natural processes that may have further affected it, for the purposes of designing an appropriate set of investigative methods for discovering any remnants of that physical evidence that might remain—albeit buried—at the site.

It is also important to note what research activities were not pursued, largely because OCA's researchers are not remotely qualified to do so. First, no attempt was made to evaluate the various proposed location(s) of the Roswell crash site. OCA relied entirely on the knowledge of the project's technical advisors and their familiarity with eyewitness reports to both choose the specific location where the investigation would be pursued, and to determine what physical evidence might be present. This is important because of the already-noted fact that no known obvious physical evidence is present today to indicate where the 1947 event took place. Second, OCA did not investigate any of the other claimed crash sites. Third, OCA made no attempt to evaluate the Roswell story and its many manifestations. Even a cursory search of the extant literature, websites, chat rooms and mailing lists will reveal that, in addition to controversy concerning the origin of and alternative explanations for the Roswell Incident itself, considerable debate exists with respect to the specifics of the location(s) of impact sites, specific crash-related events, and the parties involved. Finally, OCA was not asked to choose the project's technical advisors, nor to evaluate the accuracy of the information they provided.

For further information on the Roswell Incident, the interested reader is referred to books by one of the project's technical advisors (Randle and Schmitt 1991, 1994), as well as a series of ongoing articles by Carey and Schmitt in *Incident: The Official Magazine of the International UFO Museum and Research Center*. Schmitt and Carey also have a website (http://www.roswellinvestigator.com). Other useful references that provide insight to several aspects of the Roswell Incident include (a non-exhaustive list in alphabetical order): Berlitz and Moore (1980), Carey and Schmitt (2003), Corso (1997), Eberhert (1991), Frazier, et al. (1997), Korff (1997), Leacock (1998), Pflock (2001), Randle (1995), Saler, et al. (1997) and United States Air Force (1995). Information concerning this project, as well as a number of Roswell Incident–related links, can be found at http://www.scifi.com/ufo/ and http://www.scifi.com/roswellcrash/.

Also of interest are the July 8 and 9, 1947 issues of the *Roswell Daily Record,* which has remained the Roswell daily newspaper to this day (http://www.roswell-record.com). On July 8, the *Record* published an article with the headline "RAAF Captures Flying Saucer on Ranch in Roswell Region," with "RAAF" referring to the Roswell Army Air Field. The next day, July 9, the *Record* published several articles under the banner "Gen. Ramey Empties Roswell Saucer," including "Ramey Says Disk Is Weather Balloon," and "Harassed Rancher Who Located 'Saucer' Sorry He Told about It."

RESEARCH RATIONALE AND METHODS

The fundamental objective of the Foster Ranch site study is to determine if any remnants of the originally reported physical evidence at the site remain today. In order to accomplish this goal, a staged research strategy was implemented. The first stage entailed reviewing the physical evidence that was reportedly produced by the 1947 event. In the second, the various human and natural post-event processes and actions that could have affected the nature, distribution and detectability of that evidence were detailed. In the third, three techniques commonly used by geologists and archeologists to detect buried phenomena were applied to the site in an attempt to locate possible buried evidence. Finally, areas pinpointed using these geophysical prospection methods were targeted with standard archeological test excavations.

The research approach adopted for the investigation of the Foster Ranch site represents a form of archeological fieldwork known as *testing*. Testing strategies are used when the presence of an archeological site is known or suspected. In most cases, the presence of the site is attested to by surface evidence, usually in the form of scattered artifacts and/or suspicious mounds, charcoal stains and other non-natural surface phenomena. In such cases, testing is required if the surface evidence is insufficient to allow the archeologist to formulate definitive statements about what lies beneath the surface, how extensive it is and what methods should be used to excavate it. Testing is particularly relevant within the legal framework of cultural resource management described earlier, where such knowledge is crucial to making recommendations concerning the site's significance and ensuing decisions concerning the site's fate. In the case of the Foster Ranch site, the site is suspected, and available information suggests that physical evidence of the site may be present but buried. Thus, the present project served to test the chosen location for the presence of physical evidence, which, if found, could confirm the existence of the "site," and possibly shed some light on the event that created it.

Not surprisingly, archeologists commonly go about the business of ascertaining the nature and extent of subsurface deposits by excavating test pits whose locations are determined by the distribution of observed surface evidence. In the absence of surface evidence, they must use one or more of several methods to evaluate subsurface deposits without the costly excavation of randomly chosen test pits. One such effective (though somewhat tedious) method for evaluating subsurface deposits is soil augering. This activity involves sampling the site's space by drilling numerous holes in the ground with a bucket auger and examining the pulverized soil samples that are thus brought to the surface. Differences in the exhumed soils and the depths they come from are clues to what lies beneath. Thankfully, owing to the nature of the evidence being sought and its amenability to other, more effective methods, soil augering was not necessary in the Foster Ranch site investigation. Instead, subsurface prospection technology developed by geophysicists for the purposes of locating subsurface geological phenomena such as faults and buried water were used. The adaptation of these methods for archeological goals allows archeologists to find such things as buried architecture and other man-made phenomena.

Reported Physical Evidence and Post-Event Activities

What allegedly happened on the Foster Ranch in early July, 1947? By most accounts (see references listed earlier), on or about July 5, 1947, the ranch foreman found "metallic debris" with unusual characteristics spread across a field. He eventually was convinced to take some of the material he had collected to the Chaves County Sheriff in Roswell (Corona having no sheriff or Army Air Force Field). This act precipitated the events that became the "Roswell Incident," the story of which will not be recounted here. Accounts also indicate that a number of individuals subsequently visited the site to assess the situation. Assorted accounts indicate that witnesses saw (a) debris spread out across an area measuring ca. 300 feet wide by 0.75 miles long, and (b) a shallow, linear depression in the ground surface variously described as a "furrow" or "gouge," and measuring ca. 10 by 500 feet long, the latter presumably left by an impact of some sort (Donald Schmitt and Thomas Carey, personal communication, June 16, 2002). One eyewitness described the furrow as being up to two feet deep at the impact end, but shallower (a few inches) for most of its length (Donald Schmitt, personal communication, September 24, 2002). Finally, some accounts suggest that the concentration of debris was greatest at the presumed impact point and hence the point where the furrow was deepest (Randle and Schmitt, 1994:42).

In addition, reports indicate that, within days of the original reporting of the debris field, a large team of military personnel proceeded to the site for the purposes of removing all the debris (Randle and Schmitt 1994:37), walking, it is said, shoulder-to-shoulder across the entire debris field until all visible remnants had been acquired. Finally, all recovered debris, as well as the remains of the craft and its occupants, were allegedly removed to secret facilities for further study by the Army Air Force and/or other government agencies, which explains the lack of any material being available today for inspection or analysis.

Implications of Foster Ranch Site Characteristics
for Testing Strategies

Several aspects of this story are relevant to the research methods implemented at the Foster Ranch site investigation. Paramount among these is the fact that the two forms of physical evidence being sought by testing activities at the Foster Ranch site are (a) any residual debris not removed by the military, and (b) vestiges of the observed impact mark, "gouge," or "furrow" (the term "furrow" will be used throughout the remainder of this report). The descriptions of the debris field and furrow played a crucial role in the project's overall strategy, as well as in determining more specifically where project activities would be focused.

The Debris Field

The size of the originally reported debris field, together with the reported nature of the debris itself (light and easily moved by the wind [Randle and Schmitt, 1994:31]) and the alleged post-event cleanup, suggest the characteristics to be expected of any debris that might yet be present at or near the site.

First, the amount of debris was apparently considerable, and under the presumption that the debris was produced by partial destruction and breakup of the outer portions of whatever struck the ground, it is likely that more small pieces were produced than large ones, with the most abundant pieces being the smallest. This is because things that break usually break into more small pieces than large, a fact familiar to anyone who has found a small but painful glass sliver in a bare toe long after having cleaned up the glass debris that produced it. Thus, in terms of sheer numbers, it is reasonable to assume that the numerical bulk of the debris would have consisted of pieces that would be the hardest to find during any cleanup effort.

Second, the reported light weight of the material suggests that it may have blown considerable distances, both between the time it was originally found and the cleanup, as well as subsequently. Together, these conclusions reveal an important fact about the search for remnant debris: namely, that it is equivalent to the old aphorism "looking for a needle in a haystack." This potentially broad distribution, together with the likelihood of post-event burial by a variety of possible processes (see below), implies a "haystack" that is not only broad, but deep as well. Given reports suggesting that the debris was densest near the impact point, the Foster Ranch site testing project focused on the reported location of the furrow—in essence, looking in the densest part of the haystack. Despite the fact that this area was presumably the focus of the most intensive cleanup efforts, this strategy was deemed preferable to searching a much broader area in which the density of remnant debris would be exponentially lower.

The Effects of Natural Processes

The reported lightness of the debris also suggests that any material still present would have to have been buried or otherwise secured to the ground surface not long after the impact; otherwise, it would have eventually blown away. At least two natural processes could have contributed to burial or securing and obscuration of small, undetected debris. The first is burial by long-term natural erosion and deposition, a process known technically by geomorphologists as slope erosion or slopewash. In this process, surface sediments are moved from upslope downwards by a combination of water (which runs across the surface when the rainfall rate exceeds the rate at which the soil absorbs it) and gravity. Slope erosion will tend to move light things across the surface and eventually bury them in low areas where the eroded material accumulates.

Another burial mechanism is burrowing animals. Archeologists have long studied the effects of animal burrowing—*bioturbation,* technically—on archeological deposits. This is because the spatial relationship among excavated materials is often crucial to their interpretation, and any disturbance of the original spatial patterning, by bioturbation for example, will lead to distorted interpretations. In the desert, diurnal temperature variations, together with a general lack of vegetation large enough to hide in, means that burrowing is very common among small-medium mammals, including coyotes, foxes, rabbits and a variety of rodents. Even in the 55 years that have passed since 1947, a considerable amount of subsurface deposits have been excavated and moved to the surface, while surface materials become slowly buried. The size of the material that can be translocated is limited only by the size of the burrow. Hence, small debris could easily become buried, presumably as deeply as the burrows themselves. See Doleman (2002) for a review of bioturbation processes and Volume 17, No. 1 [January, 2002] of *Geoarcheology: An International Journal* for the latest scientific perspectives on bioturbation's effects on archeological deposits.

The Furrow

As noted above, available descriptions of the furrow suggest a linear feature measuring 10 by 500 feet, but a few inches deep for much of its length, but as deep as two feet (ca. 60 cm) along the main impact point. Descriptions of the furrow's profile are less specific, but most suggest a concave cross-section. As there is no obvious surface evidence of the furrow at present, it has presumably been either buried and/or obliterated by the same erosional and bioturbational processes described above. This fact led to the incorporation of geophysical prospection technology in the testing project. The absence of a surface-visible furrow also required that the originally observed furrow's location be determined by the project's technical advisors, one of whom had been shown the location by two separate eyewitnesses.

On the first day of the project (September 16, 2002), the OCA archeological staff met with technical advisors Schmitt and Carey for the express purpose of selecting the specific location where the testing activities would be focused. As noted above, the location was to be centered on the furrow as the area most likely to contain remnant debris, as well as being the only place to search for evidence of the furrow itself. Mr. Schmitt designated the reported initial impact point as well as the furrow's alignment (reportedly more or less straight). A permanent rebar marker was set in the ground on the same alignment, but set about 20 meters back from the initial point to allow for some error. An OCA archeologist then walked down the alignment until he was at the endpoint of the furrow as described to Schmitt by eyewitnesses. A second permanent rebar marker was set at a distance of 300 meters (985 feet) from the first marker. Despite being nearly double the 500-foot estimate for the length of the furrow, this distance was chosen to accommodate a larger fraction of the debris field. Once in place, the two rebar markers defined a baseline that served as the main axis of the coordinate system used to map the site and the locations of all geophysical prospection and archeological activities. Finally, Doleman of OCA asked how much lateral uncertainty existed concerning the furrow's location, and Mr. Carey replied that it was no more than 200 feet (61 m). Based on this estimate, the width of the study area was set at 120 meters—60 meters on either side of the designated furrow alignment—thus yielding a study area of 120 by 300 meters (36,000 square meters, 36 hectares, 15 acres—or about one-third the total area of the debris field).

Staged Research Methods

Three geophysical prospection methodologies were used in an effort to locate subsurface evidence of the furrow and debris. The research was conducted by Dave Hyndman of Sunbelt Geophysics in Albuquerque, New Mexico, with assistance from Sidney Brandwein. Detailed descriptions of the methodologies and results of all three studies are presented in a report prepared by Hyndman and submitted separately to the SCI FI Channel (Hyndman, 2002). The methods used are summarized here, while the results appear in a later section of this report.

Aerial Photograph Study

Because aerial photographs contain a wealth of information about Earth's surface that is often not readily apparent from a surface perspective, they are a commonly-used tool among Earth scientists, including geologists, biologists and archeologists. Although more properly a form of *remote sensing,* aerial photography is often used in geophysical prospection. Most aerial photographs are taken vertically by a large-format camera (9- by 9-inch negatives) from an airborne platform mounted on a specialized airplane flying parallel paths across the target area. Exposures are intentionally overlapped by about 60 percent to provide stereoscopic views. When such photographs are viewed through a stereoscope, surface topography is exaggerated and the amount of information that can be gleaned from the aerial photographs is greatly increased.

A surprising amount of aerial photography is available for most of the United States, including New Mexico, because its analytical advantages make it an invaluable tool for federal and state land management agencies, as well as scientists. Hyndman acquired aerial photographs of the Foster Ranch project area that bracketed the "event date" of July, 1947 as closely in time as possible. The acquired aerial photographs were made by the USDA Soil Conservation Service (now the Natural Resources Conservation Service) and are dated November 19, 1946 (1:31,680 scale, or one inch equal to 0.5 mile), and February 3 and 4, 1954 (1:54,000 scale, enlarged in printing to 1:27,000, or one inch equal to 0.43 mile). In addition, a single aerial photograph taken October 12, 1996 (1:40,000 scale, or one inch equal to 0.63 mile) by the United States Geological Survey's National Aerial Photography Project was also acquired for comparison with the earlier imagery.

The study of the aerial photographs served two purposes. The first was to see if any linear features that might represent a once-visible furrow were present in 1954, but not in 1946. The presence of such a feature in the 1954 photographs—but not in a 1946 image—would offer strong evidence of a furrow created in 1947, whatever its origin. The second purpose was to determine if traces of linear feature(s) observed in the 1946 and/or 1954 photography remained visible in 1996, and should suspicious linear features be present, to use the photographs to precisely establish the feature's location on the ground so that it might be further investigated.

The aerial photography from all three dates also proved valuable in addressing a number of other, unanticipated questions that arose in the course of the project.

Geophysical Prospection

Two related geophysical prospection technologies were used at the Foster Ranch site to obtain evidence of possible buried features (the furrow) and/or debris. Both acquire information on the subsurface characteristics of the soil by projecting a signal into the ground and measuring the return signal. The methods utilized—electromagnetic conductivity (EMC) and high-resolution metal detection (HRMD)— were chosen on the basis of the anticipated properties of the evidence being sought.

Electromagnetic Conductivity Survey. The EMC survey was used for the purposes of finding any buried features that might represent a buried furrow. At the heart of the method lies the fact that any furrow that was present in 1947, but which has been subsequently buried by human or natural processes, would contain deposits that are less compact and hence capable of holding more moisture than the surrounding, undisturbed and naturally consolidated sediments. These differences in consolidation or compaction would produce a higher moisture content in the feature's fill, and the higher moisture would, in turn, yield a higher electrical conductivity. In contrast, more compact soils can be expected to yield lower conductivity signals, and the limestone bedrock the lowest strength signal of all, owing to its low porosity. The EMC survey was performed with a Geonics EM-31 ground conductivity meter, which is capable of detecting patterned variations in soil conductivity. Hyndman's technical description of the EM-31 survey:

> The EM-31 is a frequency-domain electromagnetic instrument capable of detecting lateral contrasts in subsurface electrical properties to a depth of approximately 20 feet. This instrument is also sensitive to buried metal. EM-31 data were acquired approximately every 0.5 meters along traverses separated by 10 meters and oriented perpendicular to the expected trace of the furrow. Essentially, the device measures the strength of current flow induced in the substrate by a broadcast RF signal, and measuring the strength and timing of the weak return RF signal. Greater electrolytic conduction occurs in materials with higher moisture, porosity and a source of free charged particles, such as the abundant carbonates in the limestone-derived soils. (Hyndman, 2002)

In other words, the device radiates a radio frequency ("RF") signal into the ground, inducing alternating current flows with the same frequency and whose strength is determined by the soil's local conductivity. This current, in turn, generates a return radio signal which the device detects and stores in an attached data logger. The data are then downloaded from the logger at the Sunbelt offices and analyzed in a computer to yield calibrated data on spatial variations in subsurface conductivity. The spatial resolution of the EMC survey is limited by the sampling interval (0.5 meters) and the distance between transects (10 meters). The use of 10-meter transect intervals allowed the broad coverage required to survey the entire site area.

Because the EMC survey transects were tied to the grid system laid out by the archeological team and marked at regular intervals by rebar stakes and pinflags, the resulting data could be used to prepare contour maps of both the conductivity signal and the inferred electromagnetic conductivity across the site. The EMC study was conducted first and implemented across the entire 120- by 300-meter study area, including an additional 10 m buffer on each side. Once the data processing was complete (two days later), Hyndman returned to the field with the conductivity map, which revealed three to four anomalously high-conductivity locations. Two of these anomalies were elongated, located within 10 or less meters of the projected furrow, and were parallel to the furrow as determined by the project's technical advisors. This fact was a considerable surprise, as elongated high-conductivity anomalies are exactly what would be expected of a buried furrow. These two anomalies, one each located at the north and south ends of the study area, were targeted with archeological test excavations. One was also investigated with a backhoe trench.

High-Resolution Metal Detection (HRMD) Survey. The high-resolution metal detection survey was conducted for the purposes of locating any buried or obscured metallic material that might represent remains of the "metallic debris" originally reported in 1947. Almost all metallic elements are at least moderately conductive. Thus, if any metallic debris were present on or below the surface, the HRMD survey should reveal it—within the limits of the device's sensitivity. The metal detection survey was conducted using a Geonics EM-61 (HH) high-precision metal locator. Hyndman's technical description of the EM-61:

> The EM-61 is a time-domain electromagnetic instrument specifically designed for mapping buried conductors to a depth of approximately 2 meters . . . The EM-61's transmitting coil generates a magnetic pulse that penetrates the soil. This pulse induces eddy currents in the subsurface. These currents dissipate rapidly in soil, but persist in buried metal or other objects possessing a high conductivity. The long-lived eddy currents induce a signal in the receiver coil, which is integrated over a portion of the time gate between pulses. This signal, in millivolts (mV), is proportional to the quantity and depth of the buried conductor. The EM-61 (HH) provides two measurements: the 'late-time' response from strong conductors (metal objects) and the 'early-time' response from metal and highly conductive soil constituents. EM-61 (HH) data were acquired every 0.2 meters along traverses separated by 1.0 meter.

Thus, the EM-61 detects the presence of high-conductivity materials through a process that is similar to that of the EM-31 EMC meter, but which utilizes a magnetic pulse in place of a radio frequency signal. As with the EMC survey, the HRMD survey was tied into the grid system so that, once downloaded and processed, the metal detection data could be displayed in the form of a map expressed in the grid's co-ordinate system. Similarly, the spatial resolution of the HRMD survey is determined by the sampling interval (0.2 meters) and the distance between transects (1 meter). These close intervals provide a much higher spatial resolution than that of the EMC survey, but also limited the survey to an area of ca. 2025 square meters, or roughly 6 percent of the overall site area.

The HRMD survey was, of necessity, focused on parts of the site deemed promising on the basis of results from the EMC survey, and a knowledge of the site's erosion and deposits. The HRMD survey was conducted after the first two days of archeological testing, when the EMC survey results were available. Two areas were surveyed, one in the area of the most prominent electromagnetic conductivity anomaly, and the other covering the topographically low areas where deposition would be most likely to have buried any remnant debris. When avail-able after computer processing and map preparation, the results of the HRMD survey were used as the basis for placing a number of archeo-logical testing units as well as a backhoe trench in a distinct anomaly.

Archeological Activities

As noted, archeological test excavations were conducted for the pur-poses of exploring subsurface anomalies revealed by the electromag-netic conductivity and metal detection surveys, as well as to search for any buried materials of possible extraterrestrial origin that might re-main at the site. In addition to geophysical prospection anomalies, the locations of archeological test excavations were chosen on the basis of a topographically-based erosion-deposition model for the site, as well as on evidence of past, but not recent, bioturbation. The testing pro-gram incorporated standard archeological excavation methods and in-cluded photography, mapping and excavation of exploratory test pits to determine subsurface stratigraphy, as well as to investigate anom-alous localities. All excavations were documented in the files using standard excavation forms. The following archeological activities were conducted:

Grid System and Provenience Control. The OCA crew used a tran-sit, stadia rod and metric tapes to establish a datum, and a standard metric three-dimensional cartesian grid system over the site to facilitate mapping and recording (Figure 2). (As noted earlier, the location and orientation of the grid system were chosen to center on the purported alignment of the furrow in consultation with the project's technical advisors.) Locations of all major control points were recorded using averaged GPS readings (>=15 m accuracy) and marked on the appropriate USGS 7.5 min quadrangle map of the project area. Once the location and estimated axis of the target impact zone and debris field had been determined, a datum was established at the north end of the grid system and given the arbitrary coordinates 2000 North, 2000 East, elevation 100 (choice of these numbers ensures that no negative coordinate numbers will be involved, as negative numbers often produce unnecessary confusion in the field and laboratory). This datum was permanently marked with a steel rebar and stamped aluminum cap, which displays the datum's coordinates.

A baseline that follows the target area's central axis was then established using the same equipment to measure locations and relative elevations. The orientation of the grid system axis, as determined by the technical advisors, was measured at 124 degrees, 43 minutes from true north, based on a current magnetic declination of 9.6 degrees (acquired from the USGS's online geomagnetic field declination calculation software: *http://geomag.usgs.gov/frames/geomagix.htm*). Rebars and aluminum tags displaying coordinates were placed at 50-meter intervals, while pinflags and/or large nails were used to mark 10-meter intervals along the baseline. The baseline also provided control points for laying out points along the perimeters of the grid system at 10-meter intervals marked by pinflags. These pinflags, together with the marked baseline, served as control points for the EMC and HRMD geophysical prospection surveys. In addition, the baseline control points provided subdatums from which to measure and map the locations of metal detection survey localities, as well as all archeological test units and any other relevant features or artifacts.

Once the datum and baseline were established, the mapping equipment was used in an ongoing activity throughout the project to map the locations and elevations of all investigative features. A grid of points spaced at 5–20-meter intervals across the entire site was also mapped to provide input for a contour interpolation program so that the final map of the site could show the site's topography in relation to the geophysical prospection results as well as the locations of archeological test pits, backhoe trenches and any other relevant features of the site. In the laboratory, all transit data, including descriptive information, were entered into an OCA computer data base and coordinate conversion system that computed Cartesian coordinates for use in the CAD mapping system that produced the final site maps published in this report.

Test excavations. As noted above, the locations of individual archeological test units within the grid system were based on three sources of information: (a) locations of subsurface anomalies as revealed by the EMC and HRMD studies, (b) available anecdotal accounts of the locations of the impact and the resulting debris field, and (c) evaluation of the site's local topography to determine location of likely burial of former surface remains, and (4) locations of evidence for past but not recent or active animal burrowing or bioturbation. Test unit locations were measured and mapped using transit and stadia rod, as described above.

Excavations were conducted by a team of eight volunteer excavators under the supervision of OCA's principal investigator and two archeological assistants. All recovered artifacts were provenienced to a minimum of 1-meter accuracy and bagged and cataloged in the field according to standard OCA procedures. Had any human-made archeological features (prehistoric or historic) been encountered (none were), they would have been reburied and left intact for future investigations, and the test unit's location adjusted accordingly.

As part of the initial test excavations, one unit designed to determine the site's natural stratigraphy was chosen for excavation. Other test pits excavated at the same time and at opposite ends of the site served to reveal the range of variability present in the site's stratigraphy. On the basis of this, a provisional stratigraphic system was developed for use during the test excavations.

Once the results of the EMC survey were available, the locations of particular archeological test units were chosen and mapped, and test pit excavations commenced. With the exception of one series of surface-strip excavation units, all test units were 0.5 x 0.5 m pits excavated in 10-centimeter levels. Excavations were all conducted by hand using flat shovels, trowels, brushes, dustpans, buckets and tripod-mounted shaker screens (Figure 3). Vertical provenience control was maintained in terms of both 10-cm levels and natural stratigraphic units where detectable. All excavated fill was screened using 1/8-inch mesh, and soil samples of screened fill from each level were collected for possible future analysis (including microscopic inspection).

Although most of the volunteers had little or no experience in archeological excavation techniques, they were trained in the basics of excavation and the detailed record-keeping required in all archeological excavations during two intensive training sessions—one conducted at the International UFO Museum and Research Center on the evening of September 20, 2002, and the other on the first day of excavations, September 21, 2002. Each received a package of material explaining the excavation methods and rationale.

Under OCA supervision, the volunteer excavators kept detailed notes on test unit stratigraphy and any recovered artifacts using standard OCA grid excavation forms. In addition, plans and profiles of all units were photographed and/or drawn where deemed necessary. All recovered artifacts—both Native American and extraterrestrial—were entered into a field specimen catalog, with each unique horizontal and vertical provenience (e.g., grid coordinates, 5-cm level and natural stratum) receiving a unique field specimen (FS) number. Where possible, locations of specific artifacts and other features of interest within individual excavation units were measured and recorded to the nearest centimeter. Examples of standard OCA excavation forms are attached.

All recovered artifacts were collected. All Native American artifacts encountered in test excavations were placed in envelopes or bags with labels showing the excavation unit's location information and field specimen (FS) catalog number. (The FS catalog is the most important single record produced in the course of an archeological excavation, as it ties recovered materials to the specific location of their discovery—known in the archeological vernacular as "provenience" or "provenance.")

Volunteers were trained as much as possible in the recognition and distinction between natural materials and Native-American remains, and instructed to record, bag and catalog any item not identifiable as either natural or Native-American as a "Historic Material of Uncertain Origin," or "HMUO." Disposition and analysis of any suspected extraterrestrial remains were determined by pre-field consultation with BLM personnel and the project's sponsors. It is presumed that any such materials were transported to an appropriate laboratory for scientific evaluation. At the end of the project, all soil samples, artifacts, and HMUOs were transported to a secure storage facility in Roswell (see below).

All backdirt produced by excavation of test units was placed on plastic sheets next to the units, and units were backfilled at the end of the project.

Provenience Control, Unit Integrity and Security. As noted above, locations of all test units were recorded and mapped using the metric Cartesian grid system established. Within individual units the locations of recovered remains were measured vertically within 10-centimeter and natural stratigraphic levels, and horizontally with the same accuracy as the unit's location, and where possible to the nearest centimeter.

In addition, measures were implemented to ensure "provenience integrity"; that is, to avoid as much as possible the intentional or inadvertent introduction of artifacts into excavation units, either by members of the project team or unknown parties. These measures included the following:

1) Vehicular traffic was limited to existing roads and two-tracks and a central two-track that was created by driving into the site. Foot traffic was limited as much as possible to the latter two-track and specific paths at right angles to the central baseline that maximized access to test units. These measures were designed to limit disturbance of potential test locations prior to and during archeological investigation.

2) Each uncompleted excavation unit was covered with plastic and backdirt at the end of each day.

3) The bottoms of excavation units were photographed on a regular basis, including:

   a. photographing the natural surface of each test unit prior to beginning excavation to document the lack of any prior disturbance (Figure 4);

   b. cleaning and photographing the current excavation surface at the end of each day, prior to covering, and again each morning after uncovering and prior to continued excavation (any disturbance during the intervening time should be evident, Figure 5);

4) Artifact bags (coin envelopes) and small paper bags containing

sediment samples taken from test units were sealed with high-quality tape to ensure their integrity during and after transport.

Finally, all soil samples and artifactual materials recovered in the course of the test excavations were transferred to a locked security van at the end of each day. This van was sealed and kept under guard overnight. At the end of the project, the materials were transferred to lock boxes at the Wells Fargo Bank in Roswell.

Backhoe Trenching

A backhoe equipped with a specialized "archeology blade" was used to excavate trenches at the site for three purposes. The backhoe was provided and operated by Mr. Eligio Aragon of Alleycat Excavating in Los Lunas, New Mexico. Mr. Aragon—known to New Mexico archeologists as "Alley"—has over 20 years of experience using a backhoe in archeological applications. The use of exploratory backhoe trenches was not included in the original testing plan, but was approved by BLM Roswell Field Office personnel during the course of the fieldwork, once the need for such excavations was recognized.

The first use of the backhoe trenches was to search for evidence of the buried furrow by excavating a series of trenches across the furrow's presumed alignment (Figure 6). Just as subtle subsurface archeological features, such as hearths and pithouses, are revealed and easily identified in cross-sectional profiles created by excavations, it was expected that any buried furrow would similarly be evident in the sides of backhoe trenches, provided they intersected the now-buried feature.

Secondly, backhoe trenches were used to investigate the anomalies revealed by the electromagnetic conductivity and metal detection surveys—one of each—and allowed investigation of depths greater than those possible in 50- x 50-centimeter test pits. Finally, the backhoe trench provided a more comprehensive record of the site's stratigraphy, with the deeper and more elongated profiles providing a better picture of important lateral variations.

Soil-stratigraphy Study

Soil-stratigraphy is the term used by geomorphologists to refer to the study of the surface deposits that cover the landscape wherever bedrock is not exposed and which most people know as "soil." *Geomorphology* is a branch of geology devoted to the study of landscapes and their characteristic landforms. Geomorphologists understand landscapes in terms of their formation and evolution as a result of the natural processes of bedrock weathering (through which solid rock is physically and chemically broken into ever smaller fragments, eventually becoming sand, silt and clay), and the erosion, transport and deposition of the weathering products by wind and water. In fact, in soil-stratigraphy studies, an important distinction is made between *deposits* and *soils*. Deposits (or sediments) are the result of the geomorphic processes of erosion and deposition, while soils are naturally-induced changes in the deposits that take place after their deposition.

Geomorphic processes in the Foster Ranch site area are dominated by the region's geological context. The Foster Ranch site lies in the middle of a vast, undulating plateau underlain by limestone known as the San Andres Formation of the Permian geologic period and dating to about 270 million years ago. Physiographically, this plateau is known as the Upper Arroyo del Macho Basin, and consists of what geomorphologists call a *karst*—a landscape peculiar to regions of limestone bedrock. In such areas, the effect of the mild acidity of natural precipitation is to slowly dissolve the limestone's principal mineral constituent—calcium carbonate—leading over long periods of time to differential dissolution and the formation of sinkholes on the surface and subterranean cave systems such as the famous ones near Carlsbad, New Mexico (Figure 7). Because the plateau is more or less level and much of the precipitation that falls enters the ground instead of flowing across the surface, the soils in the area are classified as residual soils—that is the soils formed in deposits that are derived from the underlying bedrock and which have been transported only short distances by erosion and deposition. Other processes that contribute to the project area's deposits are (in order of importance): (a) alluvial (water-caused) processes consisting primarily of "sheetwash" in which pebble-sized and smaller particles are transported short distances across the surface when rainfall is sufficiently heavy to overwhelm the ground's ability to absorb it, and (b) blow wind which can import both sand-sized particles (which blow across the surface) and airborne dust particles (that settle out when the wind slows) from outside the area.

In contrast to deposits, *soils* form *in* deposits with stable surfaces (i.e., they are neither eroding nor aggrading) after deposition has ceased. Soils are the product of a variety of biological, chemical and physical processes on the deposit, which collectively alter the content, arrangement, chemistry and mineralogy of the original deposit, or *parent material*. In any given deposit, soils form as a vertical arrangement of related *horizons,* with the characteristics of each reflecting the processes that dominate at the horizon's depth (e.g., leaching or precipitation of certain minerals). The longer a landscape surface remains stable (usually because active sedimentation and erosion have ceased and vegetation has colonized the surface), the longer underlying deposits remain relatively undisturbed and the longer soil-forming processes operate. As a result, emerging soil characteristics become more pronounced with the passage of time. An excellent text on the principles and methods of soil-stratigraphic studies is available in Birkeland's *Soils and Geomorphology* (1999).

It is this relationship that makes soil properties useful for estimating the age of the deposits in which they formed. Such information is important to geomorphologists because the age of a landscape's soils provides minimum age estimates for the landforms on which they formed and hence offers clues to the landscape's history. Knowledge concerning a deposit's age is similarly valuable to archeologists because it allows them to infer a minimum age for the deposits in which cultural remains are being found, and hence a minimum age for the artifacts. The presence of developed soil properties also furnishes evidence that the deposits are relatively intact.

The value of the soil-stratigraphy study to the Foster Ranch project was thus threefold. First, it provided information on the natural soil-stratigraphy and estimated ages of intact strata and soils. Once the site's natural stratigraphy was documented through excavation of a "strat pit" and the earlier test pits, it was possible to determine the depth at which test excavations could be curtailed because soils—and hence deposits—older than 55 years had been reached. Second, it supplied a baseline against which to evaluate "anomalous" stratigraphy, such as might have been produced by a ground-disturbing event in the past. Finally, the study was intended to yield a better understanding of intra-site variations in stratigraphy and their implications for long-term geomorphic and soil-formation processes at the site.

RESULTS OF THE INVESTIGATION

As noted earlier, the project area lies on a broad plateau with rolling terrain. This landscape is typical of karst geological environments characterized by sinkholes and other features formed through dissolution of the underlying limestone by the mildly acidic infiltrating precipitation. This bedrock is some 270 million years old and today's plateau is the product of several million years' worth of weathering, dissolution and erosion. In addition, the ancient bedrock has been subjected to tectonic fracturing over its long existence and fracture features are present in the immediate study area. Surface deposits in the project area consist largely of in-situ weathering products, including limestone clasts and fine-grained material of a generally loamy texture, some of which is of eolian (wind-blown) origin. The Foster Ranch site lies in the northern third of this plateau, which stretches south across the Texas state line and comprises the Sacramento Section of the Basin-and-Range physiographic province (most of which lies to the west). The Pecos Valley Section of the Great Plains province lies to the east. The area's climate is transitional between the desert climates of Basin-and-Range lowlands and the better-watered Plains, with annual precipitation amounting to an average of almost 12 inches.

The project area's climate is typically continental with two-thirds of the area's rainfall being provided by the Summer monsoon rains that fall from June through September. Winters are cool, with an average January minimum temperature of 23 degrees Fahrenheit, and summers are hot, with July's average maximum daily temperature being 92 degrees Fahrenheit. Average daily maximum temperatures are above freezing throughout the year, while average minimal fall below freezing only from mid-November through mid-March. This climate supports a typical high-desert vegetation of grasses such as grama grass, and drought-tolerant shrubs such as cactus, yucca and the occasional juniper tree (Figure 2). The project area is mapped by Browne and Lowe (1994) as Plains and Great Basin Grassland.

This landscape, which stretches for miles around the Foster Ranch site, is monotonous at first glance, yet every sinkhole and swale is unique. The site chosen for the archeological testing project conducted by UNM's Office of Contract Archeology lies in an area that drains generally to the southeast (Figure 1). As noted earlier, the

120- x 300-meter grid system that defines the investigation area was centered on and aligned with the impact furrow, as reported by eyewitnesses to the project's technical advisors, Donald Schmitt and Thomas Carey. The area lies across a swale between a pair of sinkholes at the northwest end of the grid system and a bedrock high at the southeastern end (Figures 1 and 8). South of the northwestern end of the grid area lies a large sinkhole that has been truncated by erosional beveling on its east side (above truck in Figure 2), while a more intact sinkhole lies to the north (beyond people in Figure 7). Figure 8 is a view taken to the north along the grid system centerline from the bedrock high at coordinates 1718 North and 1978 East. The southeast end of the centerline lies about 20 meters (65 feet) to the photographer's right. Figure 9 is a view along the grid system centerline with the reported initial impact point in the foreground taken from coordinates 1980 North/1995 East.

Figure 10 is a topographic map of the 120- x 300-meter intensive study area based on the transit and stadia rod mapping data collected during the project. As noted earlier, the grid's coordinate system is arbitrary and grid south is oriented at an angle of 124 degrees, 43 minutes from true north, meaning that grid north is oriented approximately 56 degrees west of true north. Throughout the remainder of this report, the distinction between cardinal directions (related to true north) and grid system-based directions will be clearly indicated as follows. All coordinate specifications for the locations of various test units will be in terms of the study area grid system's coordinates, with "north" and "east" shortened to "N" and "E," respectively. As a way of simplifying locational discussions within the study area, general locations and directions will also be presented in the study area grid system. Whenever cardinal directions are required they will be noted as such. Thus, although the upper part of the study area in Figure 10 is "northwest" in relation to true north, it will be referred to simply as the "north" part of the study area. Similarly, Study Unit 2 ("SU 2" in Figure 10, whose coordinates are approximately 1924N/1985E) lies "true south" but "grid southwest" of Backhoe Trench 103 ("BHT 103" in Figure 10). Figure 10 also shows the mapped locations of archeological test pits (SUs) and backhoe trenches (BHTs) as well as a number of interesting features encountered during the project that are discussed in the following sections.

Investigations by Sunbelt Geophysics

As noted, three studies that provided crucial information to guide the archeological testing program were conducted by David Hyndman and Sidney Brandwein of Sunbelt Geophysics of Albuquerque, New Mexico. These were:

1) A study of aerial photographs from before and after July of 1947.
2) An electromagnetic conductivity survey of the entire gridded site area.
3) A high-resolution metal detection survey of selected portions of the site.

Results of all three studies were performed under contract to the SCI FI Channel and are detailed in a report delivered separately to SCI FI (Hyndman and Brandwein, 2002). The results are summarized here together with their use in archeological testing.

Aerial Photograph Study

The aerial photograph study was conducted in order to (a) determine if any linear features that might represent a once-visible furrow were present in 1954, but not in 1946, and (b) to use the photographs to precisely establish any suspicious feature's location on the ground so that it might be further investigated. As noted, the presence of such a feature in the 1954 photographs—but not in a 1946 image—would offer strong evidence of a furrow created in 1947—whatever its origin. The second purpose was to determine if traces of linear feature(s) observed in the 1946 and/or 1954 photography remained visible in a 1996 aerial photograph. It is not uncommon for features visible from the air to be undetectable on the ground, and should such suspicious linear features be present, to use the 1996 photograph to precisely establish the feature's location on the ground so that it might be further investigated.

Both the November 1946 and February 1954 aerial photography were acquired in stereo-pairs, thus allowing inspection through a magnifying stereoscope designed for just such viewing, which exaggerates topography and makes non-natural features more easily detectable. The 1996 aerial photograph was acquired in a single image. Figures 11, 12 and 13 are from Hyndman and Brandwein's report and show the 120- x 300-meter study area and immediate vicinity as depicted in the 1946, 1954 and 1996 aerial photographs, respectively. All are derived from digital scans of the actual photographs, and it should be noted that the definition and resolution in the photographs is considerably better than in the scans. The sinkholes to the north and south of the northwest end of the study area are clearly visible, as is the bedrock high in the southeast end.

No linear trace that is visible in the 1954 aerial photographs, but is not visible in the 1946 photographs, was found during inspection of the aerial photography. Thus, no evidence of a furrow-like feature that was present in 1954, but not in 1946, was detected in the aerial photograph study.

A segmented linear trace was observed within the study area that represents a southeast-trending drainage or ephemeral channel that enters the study area at the east end of the north end on a bearing of about 125 degrees from true north. In Figures 11 to 13, the drainage is a faintly darker line and area, whose appearance is caused by vegetative differences related to associated moisture variations. About a quarter of the way into the study area, the drainage turns almost due south and spreads out somewhat into the swale in the middle (compare Figures 11 to 13 with the topography in Figure 10). Observations of local bedrock suggest that the drainage's linear course in the northwestern part of the study area—as opposed to a more "natural" winding and erratic one—may be the result of underlying bedrock fracturing and topography. This is particularly true in the case of the south-trending segment, which lies between two inward-tilted blocks of bedrock that, on a larger scale, geologists would call a *syncline* (Figure 14).

A linear feature that is evident in the 1996 aerial photograph, but not in the 1946 or 1954 photographs, offers some insight into the natural processes that may have affected the site since 1947 and which may have played a role in obscuring evidence reported by eyewitnesses. The feature is an apparent vehicle track that extends from the western end of the study area almost due east and out of the area. The track is considerably less well defined than other, well-used two-tracks in the

1996 photograph, suggesting that it was not regularly used. In addition, the track was not detectable on the ground in September of 2002, six years later.

These facts shed some light on the question of how long a shallow linear feature—such as a furrow—might last in this landscape. The time required to obscure a linear depression through natural processes of erosion and re-vegetation would be a function of several factors. Not counting climate variations, the most important of these factors are (a) the orientation of the feature with respect to local slopes (approximately perpendicular), and (b) the depth and width of the original feature (assumed to be the width of a vehicle and no more than a few inches deep, although we have no way of knowing how pronounced the feature was in 1996). Given this, it seems quite possible that the shallow "furrow" described by eyewitnesses in July 1947 could have been erased in the six and a half years that passed between then and February 1954, when the "post-event" photograph was taken. Furthermore, the 55 years that passed between 1947 and 2002 may have been more than sufficient time to obscure, or even completely erase much or all of the "furrow" described by eyewitnesses. The crucial question thus becomes: how deep was the deepest portion of the furrow? As noted earlier, it may have been as much as a couple of feet deep at the initial impact point, while most of it was reportedly but a few inches deep (Donald Schmitt, personal communication, September 24, 2002). Thus, if a furrow was once present and visible, natural processes could have obscured all but a small, deep portion of it. If so, then the likelihood of detecting the feature on aerial photographs—although greater than on-the-ground inspection—would decline.

Electromagnetic Conductivity Survey

As noted, the goal of the electromagnetic conductivity (EMC) survey was to use lateral variations in the electrical conductivity of the ground to detect any buried feature that might represent a buried or obscured furrow. The instrument used—an EM-31 ground conductivity meter—is capable of detecting lateral contrasts in subsurface electrical conductivity that are largely determined by variations in substrate moisture content, which in turn is controlled by compaction and porosity. The methodology is based on the theory that, if present but buried, the furrow would produce such lateral variations in conductivity owing to the greater compaction of the soils on the sides of the furrow, particularly in contrast with the looser material that presumably filled the feature, thus obscuring it. The sides of the furrow would be more compact both because of the effects of the impact as well as the fact that older deposits become more compact and less porous over time as the result of soil-forming processes. Thus, if present, the furrow should present itself as an anomalously high-conductivity signal.

Figure 15 shows the computer-processed results of the EMC survey as a color-shaded map with the site grid system and investigative units from Figure 10 superimposed. The black dashed-line arrows show the approximate location of the drainage documented in the aerial photographs. Red shades represent the highest conductivity measurements and blue ones the lowest, with yellows and greens representing intermediate values. Almost all of the blue areas are slopes or topographic highs where limestone bedrock lies at or just below the surface. One exception to this observation is the conductivity high in the area defined by 1900–1920N/2020–2040E, just grid east of BHT 104. This high corresponds to one of the two inward-tilted blocks of bedrock that bound the drainage discussed above, being the southeastern one. In contrast, the northwestern block corresponds with the blue conductivity low at about 1940N/2020E. It seems likely that this discrepancy may be the result of the tilted and layered limestone bedrock's effects on moisture retention of the shallow overlying deposits. Under this hypothesis, moisture would be trapped by the angular uptilted edges of the southeastern block (foreground in Figure 14), but would flow down across the smooth face of the northwestern block (middle distance in Figure 14). Further study would be required to confirm this hypothesis, however. The moderately expressed conductivity highs in the northeast corner of the study area appear to represent similar cases in which near-surface bedrock may serve as moisture traps.

In addition to the above conductivity highs, three anomalous conductivity highs were observed. Anomaly 1 lies just northeast of 1920N/1980E on the northwest side of the drainage. Anomaly 2 lies in the southwest part of the study area and occupies the area defined by 1760–1780N/1940–1960E and some of the areas to the north and south. Anomaly 3 lies at the south end of the study area and is centered just southwest of 1720N/1980E. Three aspects of Anomalies 1 and 3 makes them particularly interesting. First, they are the strongest and largest of the three main anomalies. Second, Anomaly 3 is oriented almost exactly parallel to the axis of the reported furrow, while Anomaly 1 is oriented on a line between that of the drainage and the furrow. Third, each lies from 15 to 25 meters west of the centerline of the furrow, so that together the line connecting them is also nearly parallel to the reported furrow's centerline.

In contrast, Anomaly 2 lies 40–60 meters west of the supposed furrow axis, and its alignment can be projected to the northeast to connect with that of the bedrock anomaly discussed above. Given its alignment and its distance from the centerline, Anomaly 2 is assumed to have an origin similar to that of the bedrock anomaly, although this assumption was not tested.

Based on these observations, Anomalies 1 and 3 were targeted for archeological test excavations. Although three test pits (SUs) were originally established at both anomalies—one each at north and south ends and a central one—time allowed excavation of only the central one in each anomaly: SU 2 in Anomaly 1 and SU 6 in Anomaly 3. Anomaly 3 was also investigated with a backhoe trench (BHT 108).

High-Resolution Metal Detection Survey

The high-resolution metal detection (HRMD) survey was conducted on the basis of eyewitness reports that the debris found by Mack Brazel in 1947 was metallic in nature. As described earlier, the device used—an EM-61 high-precision metal locator—provides two measurements: a "late-time" response from strong conductors (metal objects), and an "early-time" response from metal and highly conductive soil constituents. The instrument's high spatial resolution (data readings every 0.2 meters along traverses 1 meter apart) means that the HRMD survey is quite time-consuming. Owing to this fact and the time constraints of a four-day project, the areas targeted for the HRMD survey were based on information gathered during the first two days of test excavations, and the likelihood that any buried debris would be most likely to occur in deposits laid down by the prevailing drainage.

The late-time results (strictly metal detection) detected only surface metal objects of known origin, including the study area grid system's baseline nails and pinflags, as well as a few pieces of rusted iron-bearing metal on the surface, all undoubtedly the product of past ranching activities (primarily cans and wire).

The early-time results, which are more sensitive and are intended to detect both metal and highly conductive soil constituents such as iron- and manganese-bearing (or ferromagnesian) minerals are presented as a color-shaded map in Figure 16. As with the EMC survey maps in Figure 15, blue colors represent the weakest signals and red ones the strongest, while greens and yellows are intermediate. The locations of centerline nails, corner nails for test pits that were present at the time of the HRMD survey, and metal pinflags marking the study area boundary are all clearly evident as red "hot spots" in Figure 16. Also present is an unmistakable feature consisting of intermediate signals that comprise a linear trace about 100 meters long and is exactly aligned with the southwest-trending segment of the area's main drainage. This trace terminates in a roughly elliptical area measuring about 30 meters by 40 meters north-south. In addition, several isolated higher-strength signals (yellow) occur along the main drainage-related trace, as well as near the center of the elliptical anomaly. Another faint linear trace also enters the HRMD-surveyed area from the east and also ends at the elliptical feature. No material or other phenomena are visible on the surface in the area of these HRMD anomalies that can account for them, indicating that whatever is causing the signal is either not visually detectable or—as seems more likely—is buried.

Both linear traces appear to correlate with possible drainages as indicated by slight U-shaped bends in the contour lines. The main trace even appears to bend to the north where the two drainage segments meet. Two conclusions seem warranted by the HRMD survey results. First, whatever material is responsible for the distinct but moderate signals appears to be correlated with local drainage patterns and thus has quite likely been transported, concentrated and buried by surface or subsurface water flow associated with the drainages. Second, the moderate signal strength could be due either to the low relative conductivity of the signal-generating material, or it is buried at some depth (assuming an inverse correlation between signal strength and depth).

At the time the HRMD survey results became available, results of other test excavations had led to a growing feeling that the lower parts of the local drainages might offer the best target for finding buried debris if any were present. The HRMD survey results buttress this conclusion considerably, and several test pits and a stripping trench (SUs 12–17) were excavated in the elliptical anomaly, as was a backhoe trench (BHT 109).

Archeological Testing

As described earlier in this report, the archeological activities conducted as part of the project constitute a methodology known as *testing*. Testing is required when the nature and extent of a site are too poorly known to allow the archeologists to develop a full excavation plan. As such, archeological testing is designed to explore a site for the purposes of evaluating its depth, extent and—in a cultural resource management context—to evaluate its significance and scientific potential. In particular, testing was required as the first activity at the Foster Ranch site because—unlike an archeological site with obvious features such as architecture to guide excavation—no obvious surface evidence exists at the site to attest to the locations of the two main reported phenomena being sought: a furrow and debris.

The aerial photography study and two geophysical prospection surveys reported on above were designed to offer clues as to potential testing locations within the area designated by the project's technical advisors as the likely impact site. Test pit locations were also chosen on the basis of an understanding of post-event processes that might have affected any originally observable evidence, including the effects of water erosion and deposition, and soil disturbance by burrowing animals (bioturbation or, strictly speaking, faunalturbation).

A total of 12 test pits, measuring 50 x 50 cm and ranging in depth from 10–90 cm, were excavated (see Figures 10 and 15 for locations of specific study units) and designated as Study Units (SUs) 2, 4, 6, 8 and 10–16 (SUs 1, 3, 5, 7 and 9 were designated and laid out, but time did not allow their excavation). In addition, two 10-m x 50-cm x 10-cm deep stripping trenches were excavated as Study Unit 17. Table 1 lists the excavated test pits under headings that reflect the rationale behind their choice, together with basic information concerning locations, dimensions, excavated levels (times 10 cm = depth) and numbers of soil samples and HMUOs recovered, plus comments.

Table 1. Excavated Archeological Study Units at the Foster Ranch Skip Site and Debris Field

Location & purpose	SW corner coordinates	Dimensions (cm)	10 cm Levels	Soil samples	HMUO bags	Comments
Stratigraphy Study Unit						
Intermediate slope, explore stratigraphy	1820N/1991 E	50 x 50	7	7	0	Revealed typical arid-land soil profile, with upper A horizon, strong cambic B horizon more orange in color than other test pits; harder and more clay towards bottom
Electromagnetic Conductivity Survey Anomaly Study Units						
EMC Anomaly 1, W of N end of furrow	1924N/1985E	50 x 50	3	4	2	Soil rocky with more clay in lower Level 3, bedrock at bottom; HMUOs from Level 1
EMC Anomaly 3, W of S end of furrow	1718N/1978E	50 x 50	5	5	2	Revealed typical arid-land soil profile, with upper A horizon, cambic B horizon, Calcium carbonate-rich Bk horizon towards bottom; HMU from Levels 1 and 2
Bioturbated Area Study Unit						
Area of past but not recent bioturbation	1824N/2001E	50 x 50	7	7	3	Surface with small limestone pieces and lots of light-colored carbonate-rich material; recent coyote burrow nearby, one possible Native-American artifact; HMUOs from Levels 4 and 6
Possible Ejecta Impact Area Study Unit						
Hillslope at distal end of furrow	1794.60N/2008.80E	50 x 50	1	1	2	Very shallow, bedrock at 8 cm; 2 HMUOs from Level 1
Prevailing Drainage Study Units						
Upper part, drainage, just E of furrow	1905N/2003.60E	50 x 50	5	5	2	Limestone clasts throughout larger with more clay in Levels 3–5; near bedrock at bottom; 2 HMUOs from Level 1

Table 1. Excavated Archeological Study Units at the Foster Ranch Skip Site and Debris Field *(cont.)*

Location & purpose	SW corner coordinates	Dimensions (cm)	10 cm Levels	Soil samples	HMUO bags	Comments
Prevailing Drainage Study Units (cont.)						
Center of drainage, W of furrow	1867.80N/1978.40E	50 x 50	5	6	1	Loamy soils, few limestone clasts, more clay towards bottom; HMUO from Level 3
Lower drainage, further W of furrow	1842.55N/1956.00E	50 x 50	5	5	0	No notes
High-Resolution Metal Detection Survey Anomaly Units						
Center of elliptical anomaly	1870N/1960E	50 x 50	5	4	0	No notes
same as above	1869N/1955E	50 x 50	3	3	0	No notes
same as above	1866N/1955E	50 x 50	5	5	4	No notes; HMUOs from Levels (1), 3 (2), 4 (1)
same as above	1866N/1952E	50 x 50	4	3	0	No notes
N-S trench across elliptical anomaly	1860-1870N/1959E 1875-1885N/1959E	2 parts, each 50 cm x 10 m	1	0	8	Pair of surface-stripping units designed to screen broad area of surface deposits
Surface Find Near SU 11	1797.70N/2006.61E	n/a	n/a	n/a	1	FS no. 31, found on surface and mapped in with transit

As discussed earlier, the acronym "HMUO" is derived from the phrase "Historic Material of Uncertain Origin"—coined in anticipation of the possibility that test excavations might recover materials produced by the reported impact that required separate cataloging and treatment. The phrase was also intended to avoid the use of overly interpretive terms. As part of their training, the volunteer excavators were shown naturally-occurring limestone pieces as well as Native-American artifacts in an attempt to limit HMUOs to materials that were unidentifiable, presumably of non-natural origin, which could not be identified as artifacts of either Native-American origin or recent historic Euro-American manufacture ("historical" is the term archeologists apply to artifacts produced by modern technology).

Finally, soil samples were collected from each level of each excavated test pit in anticipation of the possibility that the site's deposits might contain microscopic evidence of whatever occurred there in 1947. In some cases, two soil samples were taken from the same level if a stratigraphic break (i.e., an obvious change in soil characteristics) was encountered within the level.

During the four days of excavation, both the HMUOs and soil samples were kept in a sealed, locked and guarded armored security van until the end of the fieldwork. Once the test excavation phase was complete, the van, under police escort, transported the materials into Roswell, to the Wells Fargo Bank, where they were stored in a vault. As is standard procedure in all archeological excavations, all recovered materials were bagged and given a unique field specimen number and entered in the field specimen catalog. Both the bag and the catalog contain information on the material's origin—or "provenance"— expressed in terms of study unit number, study unit coordinates, and level number, thus allowing the material's original location to be accurately specified in three dimensions. At the end of each day, the recovered materials were cataloged, and the catalog was used as a checklist while the materials were transferred to the security van. A total of 55 soil samples, 25 HMUOs, and one possible Native-American artifact were recovered through archeological testing at the Foster Ranch site.

On April 15, 2003, Doleman of the UNM Office of Contract Archeology conducted an inventory of the materials stored at the bank. All materials recorded in the field specimen catalog, including HMUOs, soil samples and a possible single Native-American artifact recovered, were accounted for. During the inventory, all HMUOs were inspected, and definite or possible identifications were made. At the time, it was also discovered that latent moisture in the soil samples—which had been tightly packed into three safety deposit boxes—produced conditions perfectly conducive to the growth of molds and fungi; this, coupled with moisture-induced weakening of the paper sacks in which the samples were stored, led to disintegration of some bags and partial or total loss of nine of the 55 soil samples. Enough soil samples remain, however, to meet the needs of the soil sample analyses.

Finally, excavation of 12 test pits and two stripping trenches resulted in the removal and screening of a total of 2.375 cubic meters of deposits.

Testing Results: Stratigraphy Test Pit (SU 4)

Study Unit 4 was placed in an intermediate topographic location for the purposes of gaining an initial understanding of the Foster Ranch site's soils and stratigraphy. It also served as a test pit placed near the reported furrow some 150 meters from the reported impact point. No HMUOs were recovered. A discussion of the test pit's stratigraphy appears in the section on the site's soil-stratigraphy (see below).

Testing Results: Electromagnetic Conductivity Anomalies

Two test pits, Study Units 2 and 6, were excavated in anomalies revealed by the electromagnetic conductivity survey. SU 2 was placed in Anomaly 1, at the north end of the study area (Figure 17). The pit contained rocky soils and encountered limestone bedrock in Level 3 at a depth of less than 30 cm. Two HMUOs were recovered from Level 1. Nothing was observed in SU 2—in terms of either natural or non-natural features—that offers an obvious explanation for EMC Anomaly 1. One possibility is that the bedrock surface may be fractured and/or exhibit an uneven topography that serves as a subsurface catchment for moisture. If so, the accumulated moisture would account for the EMC anomaly as has been suggested for Anomaly 2, where tilted bedrock may have trapped subsurface moisture. In his report, Hyndman (2002) suggests "overbank" deposits as a possible source for the increased moisture content of the deposits in this area. Overbank deposits result when streams in flood "jump" their banks and deposit silt on adjacent terrace surfaces. Although SU 2 lies along the margin of the site's main drainage, there is no evidence of a true terrace. Thus, although a natural cause is possible, the origin of the EMC anomaly here remains uncertain.

Study Unit 6 was excavated on a broad topographic high at the southern end of the study area in the estimated center of EMC Anomaly 3 (Figure 8). The pit revealed a soil-stratigraphic profile fairly typical of arid lands such as the high desert of southeast New Mexico, which is discussed further below. The pit was excavated to a depth of 50 cm (5 levels), and HMUOs were recovered from Levels 1 and 2. As with SU 2, nothing was observed, either natural or non-natural, that obviously explain EMC Anomaly 3. The absence of shallow bedrock, however, came as a surprise, because it had been thought that all topographic highs in the area were underlain by bedrock. In fact, the bedrock in this location lies quite deep, a fact that may help explain EMC Anomaly 3, and which is discussed further below in the section on soil-stratigraphy.

Testing Results: Bioturbated Area

Study Unit 8 was placed in an area that exhibited evidence of past, but not recent bioturbation (Figure 18). As noted, "bioturbation" is a term for the disturbance of otherwise intact sediments that occurs as the result of plants growing (technically "floralturbation") and/or animals burrowing ("faunalturbation"). Most floralturbation occurs as the result of dead trees falling and the extensive soil displacement that accompanies the upheaving of the tree's roots. There being almost no trees in the study area, faunalturbation is the main form bioturbation takes, most notably in the form of burrowing mammals ranging from rat to coyote or fox in size (both of the latter were observed in the project area). In fact, a recently abandoned coyote or fox burrow was located near SU 8. The extensive past burrowing of the area around SU 8 was evident as a paucity of vegetation and extensive shallow deposits of lighter soils representing calcareous materials brought up from deeper deposits by the burrowers (see Figure 18). Calcareous soil horizons are a common feature of the deeper parts of arid-land soil profiles and, as such, are evidence of disturbance when they appear at the surface (see below). At the same time, the lack of open burrows suggests that the burrowing took place long enough in the past that natural processes of erosion and deposition have obscured the original burrows. Whether or not the burrowing in the SU 8 area post-dates 1947 is uncertain, but it was the opinion of Doleman that it does.

The goal of excavating a test pit in a burrowed area was based on the possibility that one or more burrows were open either at the time of the 1947 event or some time after, and that debris produced might have fallen or been dragged into a burrow. Another possibility is that surface debris was buried by later burrowing.

SU 8 was excavated to a depth of 70 cm, in part because that is the approximate depth of large-mammal burrows observed in the study area. Interestingly, HMUOs (three bags total) were recovered from levels deeper than in other test pits (Levels 4 and 6). One possible Native-American artifact was also recovered.

Testing Results: Prevailing Drainage

A total of three test pits were excavated along the prevailing drainage that crosses the north half of the study area from the northeast at ca. 1940N/2035E southwest toward 1850N/1940E and continuing outside the study area (Figure 19; see also Figure 15 and note that the arrow denoting the drainage in Figure 15 is an approximation). At its northeast end, the drainage is fairly narrow and is constrained by the bedrock "mini-syncline" depicted in Figure 14. To the southwest, it broadens out, a pattern that is visible as darker vegetation in the aerial photographs, particularly in Figures 11–12.

Test pits were excavated in the drainage under the assumption that any debris remaining from the 1947 event and subsequent cleanup might likely have been transported and buried by subsequent erosion and deposition. Such natural processes would have concentrated debris in the lowest portions of the drainage. Study Units 9, 10 and 12 were placed along the drainage from northeast to southwest, with SU 9 being adjacent to the furrow centerline (Figure 20), and SUs 10 and 12 farther to the southwest. Each was excavated to a depth of 50 cm.

Study Unit 9 revealed rocky (limestone) deposits with increasing clay at the lower levels, with evidence of bedrock being encountered at the bottom. Two HMUOs were recovered from Level 1. SU 10, in the broader part of the drainage, contained loamy deposits with few limestone clasts (rocks) and increasing clay in the lower levels. One HMUO was recovered from Level 3. No notes were made by the excavators of SU 12, also in the broader part of the drainage, although they did recover the requisite soil samples. No HMUOs were recovered from SU 12.

Testing Results: Proposed Ejecta Impact Area

One study unit—SU 11—was excavated at the request of one of the volunteers, who suggested that if a craft of some sort had in fact impacted at the north end of the study area, ejecta may have been produced that, owing to the high velocities involved, would have become imbedded in the surface "downrange." The most likely location for such secondary impacts would be the gentle slope along the north side of the topographic high at the south end of the study area. Thus, SU 11 (Figure 21) was excavated at ca. 1795N/2009E about 5 meters east of the reported furrow on the upper part of the slope (see Figures 10 and 15). The deposits here were extremely shallow (less than 10 cm), and bedrock was encountered almost immediately. Two HMUOs were recovered, however. SU 11 was also only a few meters from a surface HMUO that was recovered by the OCA staff (see below), but this proximity is coincidental, as the location of this HMUO (FS 31) was not known to the volunteer who proposed the test pit's excavation and location.

Testing Results: Metal Detection Anomalies

A total of four test pits and two shovel-stripping trenches were excavated in the area of the main anomaly revealed by the high-resolution metal detection survey (Figure 16).

These study units were the last excavated (Days 3 and 4) and the main metal detection anomaly was chosen over the drainage-aligned linear anomaly owing to the concentration of low-medium strength signals, especially since one test pit (SU 9) had already been excavated in the linear metal detection anomaly.

Study Units 13–17 were all located within 10 meters of each

other, in the eastern part of the block defined by 1860–1880N/ 1940–1960E, the area of the greatest concentration of metal detection "hits" (Figure 22). In their haste to finish excavations before the end of fieldwork, the volunteers kept little in the way of field notes. The four test pits (SUs 13–16) were excavated to depths ranging from 30–50 cm. Deposits in the test pits exhibited characteristics similar to those in Study Units 10 and 12, being loamy and having greater amounts of clay in the lower levels; none encountered bedrock. Of the four test pits, only SU 15 recovered HMUOs, which came from Levels 2 through 4.

As noted earlier, Study Unit 17 comprised a pair of shallow (10 cm), 50-cm-wide trenches excavated on the last day at the urging of two volunteers for the purposes of removing and screening large quantities of surface deposits (Figure 23). This approach was based on the realization that—with the possible exception of SU 8, which was excavated in bioturbated sediments—previous test pits all revealed apparently natural soil-stratigraphy with soil characteristics indicative of an age greater than 55 years. Barring undetected bioturbation, the deeper deposits would be unlikely to contain remains dating to 1947 or later.

Study Unit 17 was oriented north-south and divided into two trenches on the same alignment (1959E), each 10 meters long but separated by an unexcavated area 5 meters long. By itself, SU 17 resulted in the excavation of 1 cubic meter of surface deposits, or 42 percent of the total volume of deposits excavation during the project—a fact that attests to the method's productivity. This is reflected in the number of HMUOs recovered from the two trenches (8). Despite this productivity, the volumetric density of HMUOs was greater for the test pits (12.4 per cubic meter) than for the two stripping trenches (8.0 per cubic meter).

HMUOs

When HMUOs were encountered in the field, the volunteers sealed them inside labeled coin envelopes. Both the paper sacks used for the soil samples and the envelopes were prestamped with blank field specimen catalog information, including field specimen (FS), study unit and level numbers, which were written on the envelopes before sealing and entered in appropriate places on the grid excavation forms. The 25 HMUO bags transported from the field to the Wells Fargo Bank represent 17 separate FS numbers, and thus, 17 separate proveniences. As noted, all were inventoried and inspected by Doleman on April 15, 2003. Some bags contained multiple items and, in some cases, materials of more than one variety, and the 25 bags contained a total of 28 different items/materials. Table 2 lists the HMUOs, their identifications (where possible or suggested) and recommendations concerning the need for further analysis.

The 28 items/materials inspected and tentatively identified during the April 15, 2003 inventory fall into three broad material categories: natural (including mineral and bone), apparent organic (i.e., of biological origin), and of "manufactured" appearance. Six were positively identified as being natural in origin, including five mineral specimens and one bird or small mammal bone. Rock fragments found in Study Unit 15 (FS 49 and 50) might warrant further identification; although the items are clearly rocks, the apparent burning—if confirmed—might shed light on the nature of the 1947 event. The white fibers on them remain unknown.

Another three HMUOs—from Study Units 2, 8 and 9—are thought to represent naturally-occurring empty cases of metamorphosed insect larvae (pupae) and to not require further identification. The other apparently organic materials occur as three types: the thin integument-like dark green material found in Study Units 2 (FS 3, Item 1) and 17 (FS 56 and 57), the possible lichen from SU 2 (FS 3, Item 3), and collapsed sac-like items that may be small reptile eggs found in SUs 11 (FS 29, Item 1) and 17 (FS 56, Item 3). One or more examples of the dark green material warrant further analysis, as do the sac-like items, to confirm their biological origin. The lichen-like material from Study Unit 2 has probably been correctly identified. Although possibly natural, the nature of the white fibers found in Study Unit 9 (FS 23, Item 1) remains unknown and should be further identified.

The remaining HMUOs appear to be of manufactured materials, all of which should be further identified if possible. Analysis of these items should include attempts to date them if possible, perhaps by using physical or chemical characteristics to determine the date of manufacture. Further identification of the apparent shoe parts from SU

Table 2. HMUOs (Historic Materials Of Uncertain Origin) from the Foster Ranch Skip Site and Debris Field

Study Unit	Level	FS Number	Item Number	Description and Comments	Further Analysis?
Electromagnetic Conductivity Survey Anomaly Study Units					
2	1	3	1	2 pieces of organic (?) material dark green, 15–20 mm, possible algae (see also FS 56, Item 1, FS 57, Item 1)	YES
2	1	3	2	1 piece of light brown organic (?) material, ca. 20 mm, possible insect pupa case	NO
2	1	3	3	2 pieces of unknown layered organic (?) material (ca. 15 mm), possibly lichen on one side, mineral (? calcareous silt?) on other	?
6	1	5	1	1–2 pieces probable worn rubber shoe sole with molded tread (?), ca. 25 mm	YES
6	1	5	2	1–2 pieces probable shoe leather, ca. 25 mm	YES
6	2	6	1	1 piece probable worn rubber shoe sole with molded tread (?), ca. 30 mm	YES
6	2	6	2	1 piece of probable shoe leather, ca. 30 mm	YES
Bioturbated Area Study Unit					
8	4	11	1	claw-shaped limestone pebble, ca. 10 mm—natural	NO
8	6	13	1	1 piece light brown organic (?) material, ca. 20 mm, possible insect pupa case	NO
8	6	13	2	3 roughly spherical pieces calcareous mineral material, ca. 5 mm each—natural	NO
Possible Ejecta Impact Area Study Unit					
11	1	29	1	3 collapsed (?) ivory white, organic (?) items that were apparently originally hollow oblate spheres; material is thin and tough; possible reptile eggs	YES
11	1	29	2	"Flint" bagged as an HMUO is actually a natural red chert pebble, 9 mm	NO
Prevailing Drainage Study Units					
9	1	23	1	Clump of unidentified fibers, slightly "curly," 15–20 mm, pale to white in color; possibly fur, probably not nylon	YES
9	1	23	2	1 piece of light brown organic (?) material, ca. 12 mm, possible insect pupa case	NO
10	3	32	1	Flattened clump 10–15 mm of probably clothing thread (cotton?), with an unknown substance adhering	YES
High-Resolution Metal Detection Survey Anomaly Units					
15	2	49	1	Angular, non-limestone rock fragment, ca. 25 x 12 mm, with possible evidence of burning and some fine white fibers on one side—natural?	?

Study Unit	Level	FS Number	Item Number	Description and Comments	Further Analysis?
15	3	50	1	Angular, non-limestone rock fragment, ca. 30 x 25 mm, with possible evidence of burning and some fine white fibers on one side—natural?	?
15	3	50	2	Thin, curved calcareous mineral fragment, ca. 10 x 1 mm, probably a carbonate coating from a pebble	NO
15	4	51	1	10 thin (1 mm), curved calcareous mineral fragment, ca. 10–25 mm, probably carbonate coatings from pebbles	NO

Pair of 0.5 x 10 Surface-Stripping Units

17	1	56	1	Numerous small (-15 mm) pieces of dark green, "papery" organic (?) material similar to FS 3, Item 3 (see also FS 56, Item 1)	YES
17	1	56	2	Small bone fragment with articular end (probably bird or small mammal)	NO
17	1	56	3	15 mm piece of unknown organic (?) material, possibly part of a collapsed oblate "sac," tan-beige color, wrinkled, almost "leathery"	YES
17	1	57	1	3 pieces (8–15 mm) of dark green, "papery" organic (?) material similar to FS 3, Item 3 and FS 56, Item 1	YES
17	1	57	2	Fragment of apparent man-made white plastic tube (15 mm, wall thickness ca. 1.5 mm, est. outside diameter 19 mm [0.75 in]), flared at one end, dark stain on inside; probably not PVC	?
17	1	58	1	Triangular piece (ca. 18 mm) of very thin (less than .3 mm), translucent gray plastic(?), slightly shiny (between dull and glossy) on both sides; color of duct tape but lacks fibers; flexible and tough; reminiscent of gray refuse bag material	YES
17	1	59	1	2 pieces (10, 12 mm) of very thin translucent white plastic(?), rolled, split and fragile; reminiscent of plastic drop cloth material	YES

Surface Find near SU 11

n/a	n/a	31	1	2 pieces (originally 1?) of bright orange "plastic-like" material in the shape of flattened "blobs," ca. 12 and 23 mm by 5–8 mm thick; slightly flexible; surface has fine coral-like or lichen-like surface texture; fresh break is brighter and more vitreous; possibly modern plastic but otherwise unidentifiable	YES

6 (FS 5 and 6) and the "threads" from SU 10 (FS 32) might be able to determine if they are contemporaneous with the 1947 period, as well as their specific origin (i.e., military issue versus commercially available). Similarly, the thin plastic sheet-like fragments from SU 17 (FS 58 and 59) appear of man-made origin, should be confirmed through further analysis; the same is true of the "plastic tube" fragment from SU 17 (FS 57, Item 2).

Finally, the orange "blob" pieces (FS 31) that were found on the surface near Study Unit 11—although appearing to be of plastic— exhibit an amorphous oblate shape not characteristic of any man-made item known to the principal investigator. This condition might be the result of natural processes such as partial melting in the sun, or even consumption and excretion by cattle or sheep (cows in particular are known to favor the plastic material of which the "flag" part of pinflags are made). Analysis of the items' chemical composition may reveal their origin.

Possible Native-American Artifacts

The single item recovered, then tentatively identified by a volunteer excavator as a possible Native-American artifact, was identified by Doleman as being of natural origin. It is a small clast of limestone that has been chemically weathered into an odd shape. In addition, FS 11 from Study Unit 8, Level 4 was cataloged as an HMUO, but was labeled a "flint" (presumably meaning the excavator thought it was a Native-American artifact). It is, in fact, an odd-shaped limestone pebble—entirely natural in origin.

Backhoe Trenching and Soil-Stratigraphy Studies

As discussed earlier, backhoe trenching was conducted at the Foster Ranch site for three purposes. The first goal entailed excavation of trenches across the reported furrow alignment in anticipation that— if present—deeper parts of the furrow would be evident as visually detectable features that crosscut or otherwise interrupted the site's natural stratigraphy. The second goal was to investigate anomalies revealed by the two geophysical prospection surveys (electromagnetic conductivity and high-resolution metal detection) at greater depths and in a more extensive fashion than was possible with the hand-excavated test pits. Finally, the backhoe trenches provided extensive profiles for gathering soil-stratigraphic data from which to build a model of the site's natural stratigraphy and its variations. Because the soil-stratigraphy study results are based primarily on, and are important to understanding the backhoe trenches excavated in the geophysical prospection anomalies, these results are discussed first.

The methods of soil-stratigraphy are described in Birkeland's *Soils and Geomorphology* (1999) and are designed to use a variety of observed characteristics—particularly visible vertical changes in color, sediment texture and consistency (ranging from soft to hard)–to build a picture of how the sediments were deposited and under what conditions, and how subsequent soil-forming (pedogenic) processes have altered the deposits. Colors are measured using the Munsell Soil Color System, in which colors are given a three-part designation: the first denoting the hue, the second the degree of lightness or darkness (called "value," Munsell Color 1990), and the third the intensity of the hue (called "chroma"). Textural descriptions for sediments smaller than 2 mm are presented in terms of the relative proportion of the three major grain sizes: sand (0.0625–2.0 mm), silt (0.004–0.0625 mm) and clay (less than 0.004 mm; the term "clay" can also refer to a class of minerals). The presence of gravels (from 2 mm granules to 25 cm and larger boulders) is recorded separately. Consistence is measured in terms of the ease with which the intact sediments can be broken or crushed. A number of other characteristics are used in studying soil profiles as well, and all are reflective of the nature and degree of soil-forming processes that have acted on the deposit.

Soil horizons are designated in a two-part system in which the first component represents a so-called "master horizon." The "A" master horizon designates the upper zone where most biotic activity takes place, commonly giving it a dark gray or brown cast. The "B" master horizon lies below the "A," and is a complex zone or zones in which precipitation and/or leaching of chemical constituents takes place and alters the horizon's color and chemical/mineral composition. The "C" master horizon represents unaltered sediment or parent material. The second component of a soil horizon's designation describes variations within the master horizon, with various designation representing the particular processes and their characteristic results. These "subhorizon" designations most commonly apply to the B horizon. Knowledge of what soil horizons and sub-horizons are typical of the area's sediments is important, and often plays a crucial role in identifying *stratigraphic* breaks that separate past depositional units, with younger ones overlying older ones. The soil horizons developed in the units represent post-depositional periods of vegetation-controlled landscape surface stability (no erosion or deposition) that allowed soils to form. Better-developed soils are older and represent longer periods of landscape stability (see earlier "Methods" section and Birkeland, 1999). Finally, soils that show any degree of development at all are far older than 1947.

Trench in Conductivity Anomaly

One backhoe trench, designated Study Unit 108, was excavated across the southern EMC anomaly (EMC Anomaly 3) just south of the test pit SU 6 (see Figures 10 and 15). The profile, which was recorded opposite the location of Study Unit 6, revealed unexpectedly deep soils with at least two depositional units (Figure 24). The upper 20 cm comprises a poorly developed organic soil A horizon. From 20–40 cm below the surface the soil is harder and slightly redder in appearance and suggests a weakly developed "Bw" (or "cambic") horizon resulting from precipitation of iron oxides. From 40–75 cm the soil is slightly lighter and suggests a "Bk" horizon exhibiting a marked increase in calcium carbonate precipitates (note the lighter-colored zone in Figure 24). The boundary between the Bw and Bk horizons may represent a depositional boundary as well. Below a depth of 75 cm, an earlier and separate depositional unit may be represented by a slightly firmer Bk horizon. All horizons exhibit colors in the range 10YR 4.5–7/3 (dark to pale brown) and a loamy texture (roughly equal parts sand, silt and clay), although more clay is present in the lower two horizons (a not uncommon characteristic of deep soils).

Such a profile is typical of arid lands, with the lower carbonate horizon(s) being sufficiently well-developed to be in excess of 10,000 years old, and possibly much older. Two aspects of the SU 108 profile are of particular interest, however, and may be relevant to explaining the electromagnetic conductivity anomaly at this location. The first is the profile's overall depth (105 cm), a fact that demonstrates that the earlier notion—that all topographic highs in the study area represent bedrock highs—is in error. In fact, a soil auger was used to investigate the deposits below the bottom of the trench, and similar fine-grained calcareous materials were recorded to a depth of 200 cm below surface where limestone bedrock was finally encountered. The second relevant fact is the presence of four to five filled-in large-animal burrows at a depth range of ca. 65–110 cm, and all located within two meters of one another at the point where the profile was recorded. Two of these are visible in Figure 24 as darker ovals whose color reflects the fill's derivation from the darker upper A and/or Bw horizons. Each fill was noticeably softer and more porous.

If the unusual depth of the deposits in the vicinity of EMC Anomaly 3 and Study Units 6 and 108 is a localized phenomenon, then the thick accumulation might contain more moisture and account for the conductivity anomaly. The areas to the east and northeast of this area were not tested for bedrock depth, although no evidence of bedrock was noted in the far east end of the SU 108 trench, which appears to lie outside the EMC anomaly. The apparent cluster of buried, in-filled and presumably old large-animal burrows right at the peak of EMC Anomaly 3 offers a more promising explanation for the anomaly. As sizable filled-in features, they comprise a significant volume of more porous sediments that would naturally contain more moisture, and thus be capable of producing a higher conductivity reading.

No other visible phenomena that might account for conductivity Anomaly 3 were observed in Study Unit 108.

Trench in Metal Detection Anomaly

A single backhoe trench, designated Study Unit 109, was excavated across the main metal detection anomaly where Study Units 13–17 were excavated. This location is topographically the lowest area trenched and represents the bottom of the study area's prevailing drainage. The profile, which was recorded in the middle of the trench, extended to a depth of 120 cm and exhibited a soil-stratigraphic sequence that is markedly different from that of SU 108 in important ways (Figure 25). The upper 25 cm of the profile is an A horizon similar to that in SU 108. From 25–50 cm, a harder and slightly lighter-colored Bw horizon with sparse limestone gravel clasts was observed, while from 50–110 cm, a tentatively-identified harder and lighter-colored Bk horizon was recorded, also with sparse limestone clasts. At 110 cm, a possible stratigraphic break between the Bk and a softer, yet lighter-colored, older Bk horizon was noted. None of the observed horizon is as pronounced as those observed in SU 108; the Bk horizon(s) are far less well-developed and may not even warrant Bk designation (compare the light Bk horizon in Figure 24 with the slight lightening below 50 cm in Figure 25). The profile also differs from that of SU 108 in that clay is more common in all but the A horizon, and the soil colors are redder and darker overall. All hues are 7.5YR (more red and less yellow than the 10YR of SU 108), with values ranging from 4–6 (compare with 5–7 in SU 108) and chromas ranging from 2–4 (compare with 3 in SU 108).

Study Unit 9's location in the bottom of the drainage—particularly in comparison with the topographic high location of SU 108—suggests that the associated increased surface and subsurface water flow might account for the differences in the soil profiles, and possibly the metal detection anomaly as well. Subsurface water flow in particular—called "vadose flow" by geomorphologists—can play an important role in the characteristics of the soils and sediments. Although further evaluation of the local soils and laboratory testing would be required for confirmation, it is suggested that long-term vadose flow has essentially "flushed" most accumulated carbonates from the Bk horizon of the SU 109 profile, thus accounting for the considerable difference in the SU 108 and SU 109 Bk horizons. It is also suggested that the increased clay in SU 109's deposits may reflect long-term depositional conditions in which clay-sized particles are being preferentially transported, deposited and possibly translocated downwards into the deposits through a pedogenic process called "illuviation." Finally, vadose flow may have imported iron- and manganese-bearing minerals dissolved from the local limestone bedrock and concentrated them in the area of the metal detection anomaly. Red stains on exposed limestone bedrock in the study area attest to the presence of trace iron minerals in the bedrock. Similarly, clays are a complex family of minerals that consist primarily of silicon, oxygen and aluminum ("alumino-silicates"), but some also contain traces of iron and manganese. Thus, the increased amount of clay observed in the SU 109 deposits may reflect the operation of several processes linked to increased surface and subsurface water flow, and may partially explain the metal detection anomaly.

No other visible phenomena that might account for the metal detection anomaly were observed in Study Unit 109.

Trenching across the Reported Furrow Alignment

Seven backhoe trenches were excavated across the alignment of the impact furrow as reported to the project's technical advisors (Figure 9). As noted before, the centerline of the study area grid system (2000E) was aligned with the reported furrow. The seven trenches were designated Study Units 101–107 from north to south (BHTs 101–107 in Figure 10). In order to limit vehicle traffic to the existing access road that parallels the centerline (as specified in the project testing plan), backhoe trenches were split into "A" and "B" sections wherever they crossed this road (SUs 104 and 106). The trenches were excavated on the afternoon of September 24, the last day of fieldwork. The locations were chosen judgmentally, with three placed at the northern end of the alignment where initial impact occurred and the furrow was reportedly deepest; an additional four were placed along the remainder of the estimated 500-foot-long feature. Study Units 102 and 103 were placed at the estimated initial impact point, while Study Unit 103 was placed just on the other side of a limestone bedrock outcrop that eyewitnesses reportedly identified to the technical advisors (SUs 101 and 102 and the outcrop are visible in the foreground of Figure 9, while SU 103 lies just beyond the outcrop; the backhoe is visible beginning to excavate SU 104 in the distance). The location and alignment of the furrow, as well as the initial impact point originally identified by the technical advisors, was confirmed by Donald Schmitt on September 24, just prior to designation of the backhoe trenches to be excavated.

The backhoe did not arrive until about noon, and trenching did not begin until about 1 PM. During the backhoe trenching, the OCA staff were busy supervising the last of the test excavations and last-day wrap-up, as well as with the film crew and with taking photographs of study area and individual testing units and various aspects of the study area topography. Study Unit 108 was excavated first, followed by Study Units 101–109 in numerical order. Study Units 101–104A, 105, 106A, 107 and 108 were excavated from west to east, while SUs 104A and 106A were excavated from east to west. SU 109 was probably excavated from south to north.

Owing to their importance, Study Units 101–103 were inspected and photographed within an hour of excavation, while the other cross-furrow trenches were inspected during subsequent field visits to the site made by OCA personnel for the purposes of completing the transit-based topographic map and collecting. These visits took place on October 5 (Doleman only), 10–11 (OCA crew), and 16–17 (Doleman only), 2002. On October 16, the north profiles of all cross-furrow trenches (SUs 101–107) were photographed in their entirety. Only Study Unit 103 exhibited any evidence of a possible furrow, but the feature observed in the trench's south profile, about 2.5 meters east of the trench's west end—approximately 3.5 meters west of the estimated furrow alignment—provided the project's principal investigator (Doleman) with a considerable surprise. The feature was photographed immediately, but left otherwise untouched until subsequent visits.

The first photograph of the "backhoe trench anomaly" (as it came to be known), shown in Figure 26, was taken at 3:13 P.M. on September 24, 2002 with a digital camera and is unenhanced. Several aspects of the feature, which appears as an asymmetrical "V" with a rounded bottom, with the lowest point to the right of the trowel and just above the trench bottom, suggest that it marks a break or anomaly in the natural stratigraphy of the remainder of the profile. The feature is marked by a thin, intermittent line that starts below and to the right of the tip of the trowel, and rises to the left at an angle of ca. 40 degrees, continuing almost all the way to the surface. To the right of the bottom of the "V," the feature is defined by a line marking a color difference that rises to the right at a steeper angle. To the right of this line, the soil appears slightly darker and redder in the photograph. This color difference is more weakly evident to the left of the trowel, as well. Finally, the soil inside the "V" and above the trowel, where clods of grass and roots displaced by the backhoe can be seen hanging, appears to have slumped, as if the deposits inside the feature were softer and incapable of holding a vertical face such as that evident on either side of the anomaly.

These characteristics are essentially consistent with an in-filled and buried furrow or gouge resulting from an impact strong enough to significantly disturb the ground surface. The fine line defining the left side of the feature could represent either a layer of clay-dominated sediments, washed in by the first precipitation event to occur after the feature was created, or a layer of sediments compressed by the impact. In the first case, the fine sediments occur on the uphill side of the feature from which in-filling sheetwash would originate. In the second, the presence of compacted sediments on only one side might indicate that the vector of the impacting force was not vertical, but asymmetrical in a direction down and to the left. The other two characteristics that define the feature—color and soil compaction differences—are also consistent with the possibility that the feature represents a buried impact mark. Both represent apparent differences between the feature's fill and the surrounding, undisturbed and natural deposits. That the fill appears to exhibit a lighter color than the slightly darker and redder color of the undisturbed sediments to the right and left would be consistent with light-colored surface deposits washing into a depression, on either side of which the deposits retained the reddish shade characteristic of the natural soil's Bw horizon. Finally, recent fill would not be expected to exhibit the degree of compaction present in the undisturbed soils, and which result from long-term soil-forming processes.

The fact that several characteristics of the feature are consistent with expectations of a now-buried (by natural or human actions) impact mark, together with the feature's discovery within a few meters of the furrow's projected location, strongly suggest the possibility that the anomaly may, in fact, be a preserved impact mark. The feature is not evident in the north profile of the backhoe trench, however, nor in any of the other cross-furrow trenches, although it should be noted that none of the others is closer than 30–35 meters.

Alternative explanations can be offered for the anomaly in SU 103. One is that the feature is a large animal burrow (coyote or fox size) that was serendipitously bisected by the SU 103 trench. As noted earlier, a coyote or fox burrow was observed near Study Unit 8 and is assumed to be typical of such features, exhibiting a single entrance that enters the ground at an angle of ca. 40–45 degrees, the same angle as the left side of the SU 103 anomaly. In this scenario, the burrow is old enough to have collapsed and filled in, and the steeper right side of the anomaly represents a face created by the initial collapse of the deposits above the burrow. Only complete excavation of the feature can evaluate this hypothesis.

Another possibility is that the anomaly is an artifact of the trench excavation process, being either the outline of a single excavation "scoop" or "bucket wiggle," in which the side-to-side control was inadvertently moved during excavation by the backhoe operator, thus creating a bucket-shaped mark in the side of the trench. In the "scoop" scenario, the operator would remove deposits by gouging downwards, starting each "scoop" at or near the surface on the left (east) side of the feature and digging down to the right (west) at a 40–45 degree angle, then bottoming out and pulling up at a steeper, say, 60-degree angle. Two facts make this possibility extremely unlikely. First, it is known that the SU trench was excavated from west to east, not east to west. Second, the operator, Mr. Eligio (Alley) Aragon of Alleycat Excavating, is a specialist in archeological backhoe work with over 20 years experience. His standard approach is designed to limit damage to cultural features and deposits during exploratory trenching by "shaving" off a few cm of soil at a time, using a specially constructed bucket with a blade-like edge instead of teeth. Under the second possible explanation of the anomaly as a backhoe "artifact," the anomaly is viewed as a bucket-shaped "bump" mark made by the bucket moving sideways against the side of the trench. This possibility is obviated by the fact that the anomaly's side forms an angle of about 105 degrees, but the bucket's two sides are at an angle of only 90 degrees, some 15 degrees less.

Mr. Aragon was interviewed concerning the possibility that the backhoe excavation process could have somehow created the anomaly. Surprisingly, he not only said that the backhoe could not have created the feature, but that he "felt" it during the excavation; he stopped and inspected the feature at the time, observing that it did indeed contain softer sediments and confirming its existence. Despite the backhoe's power, it is, in fact, possible to "feel" changes in the consistency of materials being excavated, and Doleman has observed Mr. Aragon finding archeological features by "feel" in the past. Mr. Aragon further stated that he detected the feature in the other side of the trench and showed it to the project's still-photographer, who confirmed that—although not an expert in "dirt"—he, too, could see what Mr. Aragon was talking about.

A third possibility is that, since it occurs on a gentle slope (see Figure 6), the anomaly in Study Unit 103 might be a filled-in erosional channel that was bisected by the backhoe. This possibility seems unlikely for several reasons. First and foremost, such channels are extremely uncommon in the project area. Second, erosional channels usually have more symmetrical cross-sections than that exhibited by the trench anomaly. Finally, the fill of such channels is commonly cross-bedded fluvial (water-laid) deposits often containing well-sorted sediment layers, each of which exhibit a specific range of grain sizes (i.e., sands in some, silts and clay, or perhaps small pebble-sized gravels in others). Although the particular suite of sedimentary layers in a given channel is an idiosyncratic function of several factors, no such layers were observed in the trench anomaly (the apparent layer of fine-grained sediments lining the left side of the anomaly differs from what would be expected of true fluvial deposits). It might be argued that the absence of characteristic fluvial deposits in the channel is due to the dry channel having been filled by eolian (wind-blown) deposits, but eolian deposition plays an extremely small role in local geomorphic processes. This possibility also seems unlikely.

Because of the potential significance of the SU 103 trench anomaly, it was left undisturbed until subsequent visits to the site. The first visit was made by Doleman on October 5, while returning from an unrelated field trip. At the time, the sun was quite low and not much was visible. The site was revisited by the OCA crew again on October 10 and 11, but work efforts were focused on completing the site map and investigating the off-site linear feature discussed below. Finally, on October 16 and 17, Doleman revisited the site to complete soil-stratigraphy studies and investigate the trench anomaly again. By this time, some 22 days after its discovery, it was clear that the anomaly's defining attributes had faded considerably, with some of the fine-grained sediment line marking the left side having disappeared. It is not uncommon for certain aspects of trench profiles to fade after long exposure to the elements, especially color differences, which are enhanced in the beginning by latent soil moisture, but fade upon drying. In fact, a significant amount of rain fell on the afternoon of September 18, thus increasing the moisture content of subsurface deposits at the site.

Given the possible role of drying in the fading of the originally observed anomaly, photographs of the anomaly—both dry and after moistening with a spray bottle, and again without disturbing the face of the profile itself—were taken on October 16. Figure 27 is an unenhanced digital camera photograph of the SU 103 anomaly, taken at 4:44 P.M., after spraying the profile (the lighter soil colors to the left and bottom represent un-wetted areas). The trowel was placed in the same location as in Figure 26. The general outlines of the anomaly are still visible, with the lines defining the left and right sides occurring in the same locations as the Figure 26 photograph. The differences in soil color, albeit less pronounced, are also still present. The bottom of the anomaly appears slightly deeper, and the line defining the feature appears more uneven.

Finally, after the above photographic documentation, a trowel was used to scrape away about 2–3 cm of the exposed profile for the purpose of determining if the anomaly was a superficial feature of the profile, or if its outline continued into the deposits behind the profile. Figure 28 is an unenhanced digital camera photograph of the scraped anomaly profile, taken with a fill-in flash at 6:22 P.M. (almost two hours after Figure 27 was taken; the large nail marks the trowel's location in the previous photographs). Although the scraped profile was re-wetted for the photograph, the original outline of the anomaly largely disappeared after scraping. Hints of the color differences may be present, but the "bottom" of the feature—as defined by a pocket of lighter-colored material—is located to the left of the nail rather than the right, as in Figures 26 and 27.

The apparent failure of the SU 103 anomaly to extend into the profile suggests that, whatever its origin, the feature was, in fact, superficial. In an attempt to ascertain whether or not the color differences observed originally between the deposits inside and outside the anomaly could be independently confirmed, soil samples were taken from the SU 103 anomaly on October 17. One sample each was taken from inside, to the left and right outside the anomaly, at a depth of 35 cm below the surface (about the top of the trowel in Figures 26 and 27). The dry colors of these samples were determined via the same Munsell Soil Color System used in the soil-stratigraphy study. To ensure objectivity, the color determinations were made by another archeologist who was familiar with the Munsell system, but did not know the purpose of the measurements. Results indicate that the soils outside the anomaly exhibit a slightly more intense color than the sample from the anomaly's "fill." The sample from inside the anomaly has a color of 7.5YR 7/3 (designated "pink" in the Munsell system), while the two samples from outside are 7.5YR 7/3.5 (also "pink"), with the 3.5 chroma value indicating a slightly greater color intensity.

Overall, the origin and "reality" of the Study Unit 103 anomaly remain uncertain. The feature visible in Figure 27 is unmistakable, yet is essentially absent from the scraped profile in Figure 28. Nonetheless, the soil color differences, though slight, have been independently established, and the backhoe operator—an archeological specialist— claims to have observed the feature as well, in both sides of the trench. As noted above, the use of standard techniques of archeological excavation, in which a grid system was laid out across the feature and horizontal excavations such as those conducted in the test pits, would be required to fully evaluate the feature's existence and nature.

Other Discoveries

A number of unanticipated discoveries were made in the course of the fieldwork that bear on the project's overall goals. These include features found at the site that indicate the study area has been the focus of investigation since at least 1989. In addition, a peculiar isolated linear feature, found within about half a mile of the site, might represent a "furrow" of sorts and offers an alternative location for the 1947 events. Among other important contributions to project goals, these discoveries offer insight into the rate at which natural processes may have affected any evidence that was present in 1947 and reported by eyewitnesses, but which is not presently conspicuous.

The Debris Field "Cairn"

In the course of the fieldwork, OCA archeologists—who are trained to notice subtle surface features that inexperienced persons would likely miss—encountered several rock features on the site. These features consisted of distinctly non-natural concentrations of unmodified limestone blocks derived, undoubtedly, from nearby bedrock outcrops. The features consist of limestone rocks occurring in the form of discrete piles and closely spaced accumulations, with individual rocks in the features ranging from 15–40 cm (6–16 inches) in size. The most pronounced of these features—designated "Cairn 1" and lying at the top of the northern bedrock high at ca. 1988N/1976E—was shown to the OCA staff on September 16 by the project's technical advisors (see Figure 10). Two smaller rock features were found within a few meters of the main datum established for the study area grid system at 2000N/2000E. These are Cairn 2, located about 7 meters east of the datum, and Cairn 3, which lies just north of the datum (Figure 10). Finally, an additional rock pile was found about 20 meters north of Study Unit 2.

In an attempt to confirm independently that the area chosen in 2002 for investigation was the same as had been described in previous publications, these features were compared with a photograph of the "debris field" published in Randle and Schmitt's *The Truth about the UFO Crash at Roswell* (1994). This photograph is among photographic plates appearing in the middle of the book, between pages 168 and 169, and is stated as having been taken in January of 1990. The published photograph shows a rock feature in the foreground that closely matches Cairn 2. Figure 29 is a photograph of Cairn 2 and its setting, looking about 9 degrees west of grid north (65 degrees west of true north). The size of the rock feature in Figure 29, the number of rocks and their arrangement, and the topography and vegetation visible all match that of the photograph in the 1994 book. It thus seems likely that Cairn 2 is the same feature that appears in the 1990 photograph published by Randle and Schmitt in 1994.

CUFOS Test Pits

In the course of the initial discussions that led to the present project, the technical advisors mentioned that the Center for UFO Studies (CUFOS) had excavated a limited number of test pits at the site in 1989. These tests yielded no notable results, and apparently no written records exist. Although no specific effort was made to find evidence of the previous excavations, in addition to the various rock features described above, OCA archeologists discovered a total of eight features that consist of roughly square depressions measuring about 50 cm on a side and ranging in depth from 10–15 cm. The OCA staff quickly realized that these features might represent the 1989 test pits, and they were subsequently mapped and appear in Figure 10 as "ALN Tps" 1–8 (the "ALN Tp" designation resulted from a misunderstanding concerning the name of institution sponsoring the testing). Owing to the absence of written records of the 1989 testing, there is no way to confirm the origin of the OCA-discovered features. The features' uniformity in size and their apparent alignment along or parallel with the reported furrow centerline strongly suggests that they are, in fact, relict 1989 test pits (note the alignment of ALN Tps 1, 2 and 3 near the centerline, and the roughly parallel alignment of ALN Tps 4–8 to the west). This conclusion further confirms that the site investigated by OCA's 2002 project is the same one that has been the subject of interest since at least 1989.

Figure 30 shows one of the supposed CUFOS test pits (ALN Tp 1) and is representative of the degree of preservation evident among all eight features documented by OCA. Under the assumption that these features are in fact the remnants of 1989 test pits, of particular interest is the amount of vegetation that has grown in the features, as well as the degree of erosional degradation and in-filling that has taken place in the 13 years that passed between 1989 and 2002. Although it is not known how deep the original test pits were, and the features were not archeologically tested in the course of the present project, given their description as "limited" and "preliminary," they were probably not very deep. The amount of change affected in the features' condition sheds some light on the rate at which the natural processes of erosion, deposition and vegetation growth can affect non-natural features on the surface, in particular depressions such as the reported furrow. The presumed CUFOS test pits were by no means easily detectable, and had some not been located along the furrow alignment as designated by Schmitt, they would probably not have been discovered at all.

In the earlier discussion of the aerial photography study results, it was noted that a vehicle-caused two-track trace visible in the 1996 photograph could not be discerned on the ground some seven years later in 2002, thus implying that natural processes could erase evidence of such ground disturbance in a relatively short period of time. Similarly, the degree of naturally-induced change in the CUFOS test pits over a period of 13 years suggests that, despite the study area landscape's age, surface deposits' residual origin and the limited short-term role of erosion and deposition reinforce the possibility that shallow surface disturbances, at the least, can be obscured by these processes. Without a quantitative assessment of the observed changes, it is difficult to project the action of these processes to a period of 55 years—some five to eight times as long as those observed at the Foster Ranch site. Nonetheless, the changes documented suggest that surface features as deep as 10–20 cm might easily have become completely obscured during the 55 years that passed between 1947 and 2002.

The Alternative Furrow and the Weather Balloon

Another intriguing and potentially important discovery made by OCA personnel during the project was a curious linear feature, noted on the side of a slope bordering a small valley located about a half-mile east-southeast of the Foster Ranch site. This feature, which came to be known as the "alternative furrow," is easily visible from the two-track road that the crew used to access the Foster Ranch site every morning. The feature exhibits a number of characteristics that might be expected of the impact furrow or gouge described by eyewitnesses from 1947, while apparently not being explicable as one of a number of linear features common to the project area landscape. The possibility that it might actually represent the reported furrow warranted further investigation.

The alternative furrow is located about 100 yards south of the two-track that leads into the site and—as noted—is clearly visible from that road (Figure 31). The opportunity to explore the feature did not present itself until the last day of excavations (September 24, 2002). During the morning, while awaiting the arrival of the backhoe, the principal investigator (Doleman) parked his vehicle on the two-track and walked to the feature to inspect it. At that time he made preliminary notes, took several photographs and reconnoitered the area. When the OCA crew returned to the study area on October 10 and 11, 2002, to complete mapping of the Foster Ranch site, they also visited the alternative furrow and prepared a scaled sketch map.

The feature consists of a shallow, flat-bottomed "trench" that cuts at an average orientation of 82 degrees from true north across the gentle northeast-facing slope of the southwest side of a relatively large southeast-trending drainage (Figures 32 and 33). The sides and upper end of the feature are defined by more or less vertical erosional cuts, with the depth of the cuts declining and disappearing at the lower east end. The most distinct part of the feature measures about 32 meters (105 feet) long, although a faint depression (visible in Figure 31) continues uphill for another 25 meters (80 feet). The feature's width and depth range from about 2 meters (6.5 feet) wide and 10 cm (4 inches) deep at the east (downhill) end, to 3 meters (10 feet) wide and 36 cm (14 inches) deep at its west (uphill) end. The feature curves slightly and its orientation ranges from about 100 degrees from true north (i.e., east-southeast) at the west end to 70 degrees at the east end (east-northeast).

As noted, the bottom of the feature is essentially flat in cross-section, with a continuous covering of grass across all but the edges where the erosional cuts appear (Figure 33). The areas at both ends of the feature were inspected for any continuation that might provide clues as to the feature's origins, but, other than the 25-meter-long slight depression that extends uphill, no continuation was detectable in either direction, including the upland area some 200 meters to the west and the valley to the east.

Although the feature lacks the "U"- or "V"-shaped cross-section that might be expected from a glancing impact, its erosional edges suggest that the original cross-section might have been altered. At the same time, several aspects of the feature, including the fact that it does not continue in either direction and thus appears to be an isolated feature, suggest that it cannot be accounted for in terms of linear features either observed or that might occur elsewhere in the area, such as an erosional channel, an abandoned two-track road or a cowpath.

The rural Southwest is crossed by innumerable two-track roads that are created when rural folk drive back and forth across undisturbed ground repeatedly until the two-track is produced (cf. the incipient two-track created during the project in Figure 9). Two-tracks tend to all erode the same way as a function of the way in which they are created, with the result being two ruts and a hump in between them that continues to support vegetation. The ruts tend to erode more easily and become deeper because they have no vegetation and are subject to continuing disturbance. This pattern continues until the ruts are so deep that passing vehicles bottom out and a new 2-track is established nearby. In a few rare cases, eroded two-tracks can become true arroyos in which the rut/hump topography is completely erased, but this is clearly not the case for the alternative furrow. An abandoned two-track and its replacement were found nearby and the abandoned two-track has retained its central hump despite on-going erosion and re-vegetation. Another consideration is that well-used two-tracks tend to persist for a long time and along their entire length; however, as noted, the alternative furrow feature appears to be completely isolated within the landscape. Thus, the feature appears unlikely to represent an abandoned two-track road.

For similar reasons, the feature does not appear to be an old cow-path. While cowpaths can persist in arid landscapes for nearly as long as two-track roads, they are much narrower, usually no more than 40–50 cm (15–20 inches) wide. And, like roads, cowpaths tend to con-tinue for some distance, usually ending at a natural or man-made wa-tering facility. In fact, a pair of active cowpaths (faintly visible in Figure 31) was noted just to the south of the alternative furrow feature. In contrast to the alternative furrow, these features can be traced for a long way in both directions, with the eastern extension ending at a water facility.

Finally, the possibility of a natural erosion channel can be elimi-nated on the basis of several facts. First, such channels are quite rare in the project area, and occur only occasionally in the bottoms of major drainages. Although the entire area was searched for valley sideslopes with erosional features, none were found. The local landscape is comprised almost entirely of gentle, uneroded slopes and fairly dense grassland vegetation resulting from an estimated average annual rain-fall of 13 inches (note the absence of erosional channels in photo-graphs of the project area in this report, especially Figures 2 and 8). Even the site's prevailing drainage lacks an entrenched erosional fea-ture (see Figures 19, 20), as does the main drainage that lies at the bot-tom of the valley, east of the alternative furrow.

Second, as noted, the feature's alignment ranges from 70–100 de-grees bearing from true north, while the prevailing slope on which it lies has a bearing of 55 degrees from true north. This means the alter-native furrow runs at an angle of 15–45 degrees to the slope and to the alignment any natural erosional channel should have, provided the old dictum concerning water flow and hill direction hold true. This, to-gether with the absence of any naturally occurring side-slope ero-sional channels in the area, strongly suggests the feature is not natural

in origin. Thus, it must have been induced by a non-natural disturbance of the surface.

If common linear features can be excluded, then the alternative
furrow should be considered a non-natural feature produced by some
localized and possibly elongated disturbance of the original ground
surface sufficient to induce erosion. The feature is "non-natural" because of its non-downslope orientation and location on a slope rather
than a drainage bottom, "localized" because there is no evidence of its
continuation at either end, and "elongated" because of its dimensions
and parallel sides. Finally, the shallow uphill extension is consistent
with what might result from something "grazing" the surface and creating an elongated, parallel-sided depression that is tapered and shallowest at the ends. Erosion of the downhill portion of the feature is
in an expectable position, as erosion—or "headcutting," as it's often
called—generally starts at the bottom and works its way uphill.

These observations raise the possibility that the feature described
above represents an alternative location for the "furrow" reported by
eyewitnesses to the evidence produced by the events of July, 1947. Although the alternative furrow's average orientation of 82 degrees from
true north differs by 42 degrees from the 124-degree orientation established by the project's technical advisors, the feature does lie southeast
of the designated Foster Ranch site, along the reported path of the
"skip." In fact, the alternative furrow lies almost exactly on the line
that connects the Foster Ranch site with the "final crash site," as determined by coordinates provided to Doleman by technical advisor
Thomas Carey.

Given the alternative furrow's anomalous nature and the possibility that it might actually be evidence of the 1947 event, the aerial
photography from 1946, 1954 and 1996 that was analyzed for the purposes of finding possible "furrow" traces was re-inspected. Fortunately, all three sets of photography include the area where the
alternative furrow was found. The 1946 and 1954 photographs were
viewed through a magnifying stereoscope, while the single 1996 photograph was inspected through a magnifying glass. A feature matching
the location, orientation and extent of the alternative furrow was found
on photographs from all three dates, strongly suggesting that—whatever the nature and origin of the alternative furrow—it was present at
least as early as November, 1946, almost a year before July, 1947. It is

interesting to note, however, that the dates ("11-19-46") printed on the 1946 photograph appears in white. This fact implies that the dates were added—in the form of a light-blocking substance such as opaque ink—to the negatives after they were developed and before printing the positives (the dates on the 1954 and 1996 photographs are also white, implying a similar mode of imprinting). This, in turn, implies the possibility at least that—having been added after the negative was exposed—the dates are erroneous. This seems highly unlikely, however, and it must be concluded that the alternative furrow was already present in July of 1947.

In the course of reconnoitering the area around the alternative furrow, yet another discovery was made that—if nothing else—is ironic in the extreme, given the most common alternative explanations that have been offered for the "strange debris" found on the Foster Ranch site in 1947. The "weather balloon" explanation is probably almost as familiar to aficionados of the Roswell Incident as is the "flying saucer" one. In essence, some have argued that the "strange metallic debris" found by Mack Brazel was actually part of a secret government experiment that was later declassified and is now known as "Project Mogul." In the experiment, weather balloons were used to carry the experimental equipment aloft, including lightweight kite-like structures with reflective material that served as targets, enabling the whole assembly to be tracked by radar. The merits of this alternative theory have been exhaustively discussed and will not be reviewed here (see, among others, Saler et al., 1997; Pflock, 2001; Randle and Schmitt, 1994).

A weathered bundle of what appears to be the remains of a large balloon was found by Doleman about 100 meters southeast of the alternative furrow. Needless to say, the discovery came as a tremendous surprise. In situ, the bundle measured about 70 x 35 cm (28 x 14 inches). Several photographs were taken (Figure 34), and the entire bundle—less a few pieces that became detached—was collected and transported to Albuquerque. The material appears to be latex or rubber of some kind. The upper portions that had been exposed to the sun were quite sun-dried and fragile, while the lower, protected portions were more intact and retained considerable elasticity. Much of the material also appeared shredded. The item had been in place for some time, as attested to by the facts that (a) grass was visible growing through the material (visible in Figure 34), and (b) the degree of weathering evident on the upper, exposed portions.

The item was taken to the National Weather Service (NWS) office in Albuquerque, where it was identified as definitely being a high-altitude payload balloon of the sort used by the NWS to carry weather-monitoring instruments aloft, but which are also used by a variety of other government agencies and scientific institutions to transport high-altitude experiments and equipment. NWS personnel also noted that the specimen's partly shredded condition is consistent with what happens when weather balloons and other high-altitude payload balloons reach great heights, expand to the point where their elastic limit is exceeded, and burst (during this process, the balloons increase from about six feet in diameter to house size). They also showed Doleman one of the balloons currently in use by NWS-Albuquerque, and the materials are clearly quite similar. Finally, they hazarded a guess that the balloon found near the alternative furrow is probably not more than about 10 years old, although they expressed some uncertainty and suggested that the manufacturer (NWS balloons are currently made by Kaysam Inc.) might be able to better date the balloon. They also noted that typical payloads for such balloons are about eight ounces (half a pound), and that such a lightweight payload could hardly create a "furrow" of any sort.

Thus, the weather balloon found near the alternative furrow is probably an unusually ironic coincidence, and not the wreckage of whatever created the furrow in the first place. Nonetheless, further analysis and dating of the balloon would be required to confirm this conclusion.

SUMMARY AND RECOMMENDATIONS

Geophysical prospection and archeological testing were conducted at the Foster Ranch skip site and debris field to determine if any physical evidence remains of the impact event that reportedly took place there in July, 1947. The two kinds of evidence reported by eyewitnesses are unusual "metallic debris," and a "furrow" or "gouge," both of which attest to the severity of the impact. The impact was apparently of a glancing nature, as no evidence of the impactor itself was reported at the site, and at least two other "final crash sites" have been reported at some distance to the southeast (Randle and Schmitt, 1994). The project's methodology qualifies as archeological *testing* rather than full *excavation,* because the investigation was limited in scope and designed to evaluate the site's potential, rather than to examine phenomena known from visible surface evidence, which is lacking at the site. The specific location chosen for the investigation was determined by the project's technical advisors, Donald Schmitt and Thomas Carey, and was based on eyewitness reports.

Owing to the present-day lack of surface evidence, a staged research approach was adopted, in which geophysical prospection methods commonly employed in archeological exploration were used, together with a knowledge of how natural processes might have buried former evidence, to guide placement of subsequent archeological test excavation units and backhoe trenches. All on-site investigations were tied into a grid system established at the site, with the centerline aligned with the furrow, as reported to the technical advisors. Research began with a study of aerial photographs of the project area taken in 1946, 1954 and 1996. The photographs were studied to determine if linear traces of the furrow could be detected, particularly traces that appear in post-1947 photographs but are absent from the 1946 photography. A segmented, two-part linear trace was found in the northern end of the study area that undoubtedly represents the vegetative signature of the site's prevailing drainage. The photography also proved valuable in assessing the age of a furrow-like feature that was found on the ground, about one half-mile southeast of the study area.

Two geophysical prospection technologies were used at the site. One, an electromagnetic conductivity (EMC) survey that is sensitive to variations in ground moisture, was conducted across the entire site to detect anomalously high ground conductivity that might represent a buried furrow. Four such anomalies were detected, two of which—Anomalies 1 at the north end, and 3 at the south end of the site—are aligned with the reported furrow. Archeological test pits (Study Units 2 and 6, respectively) were excavated at Anomalies 1 and 3, and a backhoe trench (SU 108) was excavated across Anomaly 3. No evidence of a furrow was found at either anomaly. Deposits at Anomaly 1 are quite shallow, and although the anomaly remains unexplained, it is possible that the shallow bedrock's topography is acting as a moisture trap that yields high conductivity readings. At Anomaly 3, deposits are unexpectedly deep, possibly representing a bedrock depression that yielded similarly high conductivity readings. Anomaly 3 also corresponds with a cluster of large, buried animal burrows (found in the backhoe trench) whose relatively porous fill likely contains greater moisture that could also account for it.

A high-resolution metal detection survey was also conducted for the purpose of detecting buried or obscured "metallic debris" left over from the impact and subsequent site cleanup that was reportedly conducted by U.S. military personnel. The metal detection survey focused on the site's prevailing drainage under the hypothesis that any such debris would have been concentrated there by natural erosional processes during the 55 years that had passed since 1947. A continuous linear trace of moderate to strong metal detection readings was detected along the southwest-trending segment of the drainage. The trace culminates in an elliptical area of similar signals that measures about 30 meters by 40 meters north-south in the west-central portion of the study area. A fainter linear trace that corresponds to a smaller drainage also converges at the elliptical metal detection anomaly.

Three archeological test pits (SUs 9, 10 and 12) were placed along the drainage, and four (SU 13–16) were excavated in the elliptical anomaly along with two 0.5- by 10-meter stripping trenches and a backhoe trench (SU 109). No obvious metallic materials were encountered during excavation of any of the drainage-focused or metal detection anomaly test pits, or in the backhoe trench. Soil-stratigraphic studies conducted as part of the project, however, revealed important differences between soil profiles on topographic highs and lows that may explain the metal detection anomaly. Long-term erosion and deposition processes and/or subsurface (vadose) water flow may be concentrating ferromagnesian minerals and/or clays derived from weathering of the local limestone bedrock along the prevailing drainage.

In the course of the archeological test excavations items, not clearly identifiable as of natural, Native-American or industrial manufacture, were cataloged as "HMUOs" ("Historic Materials of Uncertain Origin") and saved for future analysis and identification. A total of 25 HMUO bags from 17 separate proveniences—comprising 28 individual specimens (items or materials)—were recovered, as were 55 soil samples. Two items were identified by volunteer excavators as of Native-American origin, but both were subsequently identified as natural. Both the HMUOs and soil samples (see below) were kept in a bank vault in Roswell pending retrieval for various analyses.

With the exception of the "fibers" found in SU 9, the 28 individual HMUOs fall into three broad material categories: (a) definitely natural (including five mineral specimens and one bone), (b) unidentified but of probable organic (i.e., biological) origin, and (c) of apparent manufactured origin. The definitely natural items do not require further analysis, nor do some probable organic items such as the insect pupae. The other apparently organic materials, including the thin dark green material (three examples), the possible lichen (one example) and the sac-like items (two examples), warrant further identification, as do the SU 9 fibers, which may be of organic or manufactured origin. Finally, all of the HMUOs tentatively identified as "manufactured," including rubber and leather shoe parts (several), thin plastic sheet fragments (two), plastic tube (one), apparent threads (possibly cotton) and the "orange blob" pieces should be further identified through appropriate forensic analyses.

Forensic analyses of the HMUOs should be conducted for two purposes. The first is to confirm, if possible, that the items are, in fact, of terrestrial and identifiable origin, whether biological, mineral or manufactured. For those that appear to be manufactured, attempts should be made to both determine their specific origin and use, as well as to date them, if possible, by identifying diagnostic period-of-manufacture characteristics. Identification of any manufactured items that date to the 1947 era may shed light on the events that took place at Foster Ranch at that time.

In general, the term "forensic" refers to the use of scientific methods in legal issues and courts of law. Commonly, forensic analytical methods are applied to material evidence recovered from crime scenes for the purposes of identifying the material and elucidating the events that took place, and the parties involved. Forensic materials analyses fall into two general categories. The first might be called specialist-based analyses, in which a scientist who specializes in a particular kind of material identifies specimens, often consulting a library of representative examples. For example, a forensic entomologist specializes in identifying insects and insect-related items, and often uses his knowledge of insect behavior and ecology to adduce facts about crime scenes where insect evidence is found. The second category might be called chemistry/physics-based and involves the use of a variety of analytical technologies, to determine the chemical or elemental composition of unknown materials and use the resulting information to identify them. One such analysis is gas chromatography, which separates a material's chemical constituent on the basis of their molecular weight. Both approaches may well be required to identify the HMUOs from the Foster Ranch site.

During the testing, a total of 55 soil samples were collected, one from each 10-cm level of most excavated test pits. The samples were collected in anticipation that any microscopic debris present could be detected by specialized laboratory analyses. Samples of identified soil horizons were also collected during the soil-stratigraphy study of the backhoe trenches. A representative sample of the test pit soil samples should be analyzed to determine if they contain unusual chemical constituents not consistent with the composition of the project area's naturally-occurring soils. Suggested analytical techniques include X-ray fluorescence (XRF) or inductively coupled plasma mass spectrometry (ICP-MS). Both techniques measure the elemental composition of a sample, including even very small amounts. ICP-MS is particularly sensitive and can detect traces of almost all known elements. If microscopic debris produced by the reported 1947 impact is present in the soil samples, it should be revealed by such analyses unless its composition was identical to that of local soils—a highly unlikely proposition. Comparison with analyses of a control sample of similar soils taken from an off-site location would serve to bolster any analytical results.

Mineralogical analysis of the soil horizon samples taken during the soil-stratigraphy study should be conducted to determine if mineralogical differences between high-topography (SU 108) and drainage-bottom (SU 109) soils can account for the metal detection anomaly found in the site's prevailing drainage. X-ray diffraction analysis is the standard method used by geologists and soil scientists to determine mineralogical content. The method is particularly suited to detailing the suite of clay minerals present, and should be valuable in assessing the hypothesized role of clays in general, and ferromagnesian minerals in particular, in the metal detection anomaly.

Backhoe trenches, excavated by an operator with extensive experience in archeological settings, were used in the testing project for three purposes. The first was to investigate the anomalies revealed by the electromagnetic conductivity and metal detection surveys more extensively than is possible via archeological test pits. The second was to provide profiles for the soil-stratigraphy study. Trenches excavated across the most interesting EMC anomaly and the main metal detection anomaly yielded no direct evidence to explain either. Soil-stratigraphic information from these two profiles, however, suggests possible explanations for both. The EMC anomaly may have been produced by excess ground moisture accumulated in an unusually deep bedrock depression, and/or a cluster of large animal burrows whose location coincides exactly with the center of the anomaly. Comparison of the documented SU 108 and SU 109 soil profiles suggests a possible explanation for the metal detection anomaly at SU 109. SU 108 is located on a topographic high and contains a well-developed calcium carbonate soil horizon. The profile of SU 109, located along the site's prevailing drainage, lacks a carbonate horizon, but contains significantly greater amounts of clay. The absence of a carbonate horizon in SU 109 may have resulted from long-term dissolution and removal by vadose zone (subsurface) moisture through-flow that is also concentrating iron- and manganese-bearing clay minerals and/or ferromagnesian minerals weathered from the local limestone bedrock. Mineralogical analyses of soil samples collected from these profiles should serve to evaluate this hypothesis.

Finally, seven backhoe trenches (SUs 101–107) were excavated across the reported furrow alignment for the purposes of detecting buried evidence of the feature. One trench—Study Unit 103—revealed a "V"-shaped stratigraphic anomaly on the south side of the trench, within a few meters of the projected alignment. Although the feature was quite distinct upon its original discovery, it faded within a few weeks, owing (in part at least) to drying. The anomaly is all but invisible in a cleaned profile, but the backhoe operator, whose experience in archeological settings is legendary in New Mexico, claims to have noted and inspected the feature as well. The anomaly was originally identified on the basis of a thin sediment line and color differences between the feature's "fill" and surrounding deposits. Comparison of the "fill" with adjacent deposits using the Munsell Color System indicates a slight color difference. Further investigation of the feature is recommended, including a full excavation using standard archeological field methods.

In addition to the results of the designed research, three unanticipated discoveries were made that bear on the project's research focus. The first of these is eight apparent test pit remnants that were probably excavated as part of a CUFOS-sponsored testing project in 1989. Their presence attests to the chosen sites having not changed since at least 1989. At the same time, the test pit's partially filled-in and vegetated condition offers valuable insight into the rates at which the natural processes of erosion, deposition and re-vegetation can erase surface features on the high desert landscape, and suggests that the reported furrow could possibly have been entirely erased by natural processes in the 55 years that passed between 1947 and 2002.

Two other discoveries are a linear, furrow-like feature found on a hillslope about one half-mile southeast of the Foster Ranch site, and a weather balloon found about 100 meters to the southeast. For a variety of reasons, the feature—termed the "alternative furrow"—does not appear to represent any linear feature common in the area, including an abandoned two-track road, a cowpath or even an entrenched erosional channel (which are, in fact, extremely rare). The feature appears on the aerial photographs used in the research, including those from 1946, indicating—unless the photographs' dates are erroneous—that the feature was present in November, 1946. Personnel from the National Weather Service's Albuquerque office confirmed the balloon's purpose as carrying high-altitude payloads, and estimated its age at no more than 10 years. Given that the reported furrow was one of the two kinds of physical evidence being sought by the project, acquisition and study of additional aerial photography of the area seems warranted. Similarly, given the role of "weather balloons" in alternative explanations of the Roswell Incident, further analysis of the balloon is recommended for the purposes of shedding more light on the balloon's origins, and its relationship—if any—to the alternative furrow.

The Foster Ranch archeological testing project represents the first comprehensive attempt to locate extant physical evidence of whatever event occurred on the Foster Ranch site in July, 1947. The project sought to uncover remnants of the two most commonly reported kinds of physical evidence: a furrow and unusual debris. No conclusive evidence of either was found, but two furrow candidates were discovered and should be further investigated. In addition, a number of unidentified items warranting forensic analysis were also recovered.

Finally, soil samples collected during the project should be subjected to specialized analyses for two purposes: (a) to better understand the electromagnetic conductivity and metal detection anomalies, and perhaps more importantly, (b) to determine if microscopic evidence of debris is present in the form of traces of unusual elemental or chemical composition.

REFERENCES

Benson, Saler, Charles A. Ziegler, and Charles B. Moore. 1997. *UFO Crash at Roswell: The Genesis of a Modern Myth.* Washington, D.C.: Smithsonian Institution Press.

Berlitz, Charles, and William L. Moore. 1980. *The Roswell Incident.* New York: G.P. Putnam's Sons. (Also published in 1980 by Berkeley Publishing Group, New York.)

Birkeland, Peter W. 1999. *Soils and Geomorphology,* 3rd ed. New York: Oxford University Press.

Brown, David E., and Charles H. Lowe. 1994. *Biotic Communities of the Southwest* (map). Salt Lake City: University of Utah Press.

Carey, Thomas J., and Donald R. Schmitt. 2003. *Witness to Roswell: The Most Comprehensive Investigation Ever Made into the 1947 Roswell Incident.* Roswell: The Triton Group.

Corso, Philip J. 1997. *The Day after Roswell.* New York: Pocket Books.

Eberhert, George M. (editor). 1991. *The Roswell Report: A Historical Perspective.* Chicago: J. Allen Hynek Center for UFO Studies.

Frazier, Kendrick, Barry Karr, and Joe Nickell (editors). 1997. *The UFO Invasion: The Roswell Incident, Alien Abductions, and Government Coverups.* Amherst: Prometheus Books.

Hyndman, David A., and Sidney S. Brandwein. 2002. *Geophysical Investigations of a Suspected UFO Crash Site Near Corona, New Mexico.* Albuquerque: Sunbelt Geophysics.

Korff, Kal K. 1997. *The Roswell UFO Crash: What They Don't Want You to Know.* Amherst: Prometheus Books.

Leacock, C. P. 1998. *Roswell: Have You Wondered? Understanding the Evidence for UFOs at the International UFO Museum and Research Center.* Ann Arbor: Novel Writing Publisher.

Munsell Color. 1990. *Munsell Soil Color Charts,* 1990 ed. (revised). Baltimore: Munsell Color Macbeth divisions of Kollmorgen Instruments Corporation.

Pflock, Karl T. 2001. *Roswell: Inconvenient Facts and the Will to Believe.* Amherst: Prometheus Books.

Randle, Kevin D. 1995. *Roswell UFO Crash Update: Exposing the Military Cover-up of the Century.* New Brunswick: Global Communications.

Randle, Kevin D., and Donald R. Schmitt. 1991. *UFO Crash at Roswell.* New York: Avon Books.

Randle, Kevin D., and Donald R. Schmitt. 1994. *The Truth about the UFO Crash at Roswell.* New York: Avon Books.

United States Air Force. 1995. *The Roswell Report: Fact versus Fiction in the New Mexico Desert.* Washington, D.C.: U.S. Government Printing Office, Superintendent of Documents.

....

SCI FI and OCA have been in contact with Governor Richardson's office, and are currently working out details that might take the investigation to its next logical stage. But for now, the mystery continues. No "Case Closed," no simple yes or no answers. After nineteen months of planning and strict scientific analyses, only two facts are immutable: (1) something happened out at the J. B. Foster Ranch in July 1947, and (2) we must dig further—figuratively and literally—if we are ever to ascertain, once and for all, what that "something" was. Until then, our minds must remain open to the possibilities . . . and to the enduring speculation that is the Roswell Incident.

AFTERWORD: THE TRUTH BE TOLD

Every single day, the line between science fiction and science fact blurs a bit more. In fact, many things that used to exist solely in the realm of fiction are now fact. We have been to the moon; we have cloned animals (and possibly humans); and in the more fortunate nations, nearly anyone who wants a computer has one. Entertainment is piped into our homes twenty-four/seven. Our keys, watches, exercise equipment, cars—even our clothing—have "smart" technology. The cell phone is Captain Kirk's communicator; the PDA, Mr. Spock's tricorder. Advances in physics have changed the way we look at the world, from the smallest quantum particles to galaxies on the other side of the universe.

Viewers of the SCI FI Channel want us to explore this blurry line, and those who work at SCI FI are happy to oblige. That is why we funded the 2002 scientific dig at Roswell, and partnered with the University of New Mexico to search for possible physical evidence left behind at the site in 1947. It is also why we are now investigating the reported UFO incidents in Kecksburg, Pennsylvania, and Rendlesham Forest in Great Britain. We are also looking into other areas of science fiction/fact, including the seeming convergence of man and machine, explored in the *Matrix* and *Terminator* movies, and in SCI FI's own *Battlestar Galactica* miniseries. We believe that humanity must look very carefully at the benefits and dangers of such advanced science as artificial intelligence and nanotechnology, and that these relatively new fields need to be regulated carefully.

Reporters and colleagues have asked me why SCI FI is putting its corporate weight behind such initiatives. After all, we are traditionally taught that corporate action is motivated by only one thing: the pursuit of profit. In the television business, that means the pursuit of Nielsen ratings, and even though our investigative and UFO specials do earn strong ratings, there's much more to them than that.

While the profit motive is always important for any corporation, over-all, corporations are just a collection of individuals, each with personal interests of his or her own—interests that go beyond pure profit. As a result, most large companies sponsor community and national initiatives, which often receive overwhelming support from employees. This enthusiasm creates a sort of "magic," out of which grows charitable support, scholarships, and political activism. Employees become more motivated and feel more positive about the company for which they work. It's an old axiom, but it still holds true—companies do well when their people are allowed to do *good*.

In the television business, extracurricular activities of this kind are often linked to the particular network's "brand." For example, cable music network VH-1 sponsors a politically oriented initiative called "Save the Music," which encourages the restarting of music education programs in high schools across the country. Discovery's *Animal Planet* helps to support endangered species. And MTV's "Rock the Vote" campaign encourages its viewers to participate in the democratic process. All of these projects are completely appropriate for each network's brand, its viewing audience, and its overall corporate agenda.

SCI FI Channel's employees also want to make a difference, so in early 2002, a group of us began seriously to explore what initiative could be launched that would be appropriate to the spirit of the network. We knew that SCI FI—which can now be seen in over 83 million U.S. homes (reaching more than three-fourths of the nation's population)—could really have an impact. Finally, after many brainstorming sessions searching for the right idea, the answer became clear—SCI FI would push for the truth behind all of the many reports of UFOs and other unexplained phenomena. In our opinion, no other company was better suited to ask these questions.

We were excited, since we knew we were the first (and so far, only) company ever to pursue such an initiative. We decided to attack this issue on a number of fronts, both on-air and -off. First, we would encourage and, when possible, fund scientific study into these phenomena. And, over the last year, we have succeeded in doing so. From the dig in Roswell, to three separate scientific symposiums, to lab work studying HMUOs, SCI FI is putting its money behind legitimate scientific investigation.

Second, SCI FI is attacking this issue on a political level. Those of us at SCI FI who are involved with the initiative passionately believe that unless current national security is at stake, there is no apparent reason for the government not to release its complete files on UFO cases. As a matter of public policy in all matters, UFO-related and otherwise, we believe that government should be as "transparent" as possible with its citizens. And in the case of UFO reports, we believe that the public has a right to know what's going on in the skies above us. Polls show that even skeptics believe these questions should be asked.

Regarding the 1947 incident outside of Roswell, New Mexico, there seems to be no conceivable reason the government cannot release all of its files, complete and uncensored, concerning this event. After all, what national security issue can be at stake regarding something that happened in 1947? The Cold War with the Soviet Union is over, and Russia is considered a friendly nation. If the Roswell Incident truly is nothing more than the crash of a then-secret military project, which the government has so often claimed, then the American public should be shown the files, complete and unexpurgated. And if an extraterrestrial craft did crash at Roswell, then we should be told. We do have the right to know.

Personally, having been to Roswell, and now having looked deeply into other UFO cases, I feel that something is going on out there. Taken together, too many reports come in every single year to be all dismissed casually as most governments try to do. What that "something" is, I don't yet know, but I am motivated to find out.

As previously mentioned, Roswell is not the only case that the SCI FI Channel is investigating. Another lesser-known case, but one whose trail is warmer, happened in Kecksburg, Pennsylvania, in 1965. Something crashed in the woods on the outskirts of this small town one December evening. Townspeople, as well as a number of reporters, went out to investigate. Witnesses claim they saw an acorn-shaped metal object with "odd" symbols on it. Within hours, a caravan of military vehicles and personnel showed up in town, whisked the item away, and in very strong terms, told a number of locals to keep quiet about the incident.

Because this case is more recent, many of the witnesses are still alive, and are now talking about what they saw. SCI FI is supporting a Freedom of Information Act request for the release of the full and uncensored reports on this incident. If, as some claim, the object that crashed outside Kecksburg was the remains of a Soviet spy satellite, then we should be told. The Cold War is over, and any technology the United States captured from this satellite would be long obsolete by now anyway, so where would the harm be from a "national security" point of view? However, if, as many witnesses claim, the object that crashed outside Kecksburg was not of our world, then humanity clearly has the right to know the truth. That kind of information should not be kept in the hands of only a relative few at various government agencies and military departments. That would be just plain wrong.

Another even more recent case that has caught SCI FI's attention happened in and around two NATO military bases in Great Britain's Rendlesham Forest in 1980. A craft of some sort reportedly touched down around the base numerous times, and was seen by both officers and enlisted personnel. These men were reportedly ordered not to speak about what they had seen. Now, however, many are coming forward with a story that is at once startling, compelling, and chilling, especially because this sighting occurred at a military base where nuclear weapons and technology were reportedly located.

As you've read in this book, SCI FI has retained the Washington, D.C., lobbying firm of PodestaMattoon to help us fight for the truth at the highest levels of government. We have also received vocal support from John Podesta, President Clinton's former chief of staff, who brings with him unquestioned credibility and an intimate knowledge of how Washington works. Mr. Podesta has told SCI FI that freedom of information and transparency in government were very important to the Clinton administration. Now, in the post-Clinton years, Mr. Podesta is dedicated to continue working for truth and openness, whenever possible, in government. His belief in the Freedom of Information Act, his outspokenness on UFO and advanced technology issues, and his support for SCI FI's efforts have been invaluable to our ongoing fights.

While I write this, it's been over a year since my trip to Roswell, New Mexico. The images and memories I have from that extended weekend are still vivid in my mind. I feel so fortunate to have been a part of this initiative, as well as our new projects.

Clearly, many things are going on in our world that cannot be easily explained. Polls show that a majority of Americans believe that the government is covering up information on UFOs. Additionally, more and more scientists believe that it is highly unlikely that Earth is the only planet with life on it. Will SCI FI's initiatives one day definitely prove that aliens have visited Earth? Maybe, maybe not. Ultimately, success in this initiative won't be defined by actually finding a UFO. Success will come when the full stories of all of these incidents are released to the public, and the truth is known, whatever that truth might be.

If you've read this book, then you are someone who is clearly interested in these issues. SCI FI Channel has opened these doors again, but now it's up to motivated individuals around the world to push their way through. Whatever the reasons or answers, it's important for the truth to be told, and for science to be served. So keep reading, keep watching, and never stop asking questions or pressing for answers. Truth is everyone's responsibility.

Thomas P. Vitale
Senior Vice President, Programming, SCI FI Channel
January 2004

Check in with www.freedomofinfo.org for updates on the SCI FI–supported FOIA initiatives. Also, log on to SCIFI.COM for information on SCI FI's upcoming documentaries, which explore the line between science fact and science fiction.

POSTSCRIPT

In July 2003, SCI FI sponsored the installation of a memorial marker at the crash site near Roswell, New Mexico, to honor those who came forward to report what they witnessed more than half a century ago. Among those in attendance were representatives from the International UFO Museum and Research Center in Roswell, the SCI FI Channel, UFO investigators, and descendants of Mack Brazel.

The marker reads:

In July of the year 1947 a craft of unknown origin
spread debris over this site. Witnesses would report
materials of unearthly nature.

In September of the year 2002 the SCI FI Channel
brought scientists from the University of New Mexico to
search this ground for evidence of that fateful night.

Be it observed, that whatever the true nature of what
has respectfully become known as the Roswell Incident,
humankind has been forever drawn to the stars.
Dedicated July 5, 2003.

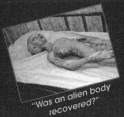